EXERCISE and STRESS RESPONSE
The Role of Stress Proteins

CRC Series in Exercise Physiology

Series Editor
Ira Wolinsky

Published Titles

CONCEPTS in FITNESS PROGRAMMING
Robert G. McMurray

EXERCISE and DISEASE MANAGEMENT
Brian C. Leutholtz and Ignacio Ripoll

PHYSIQUE, FITNESS, and PERFORMANCE
Thomas Battinelli

MUSCULOSKELETAL FATIGUE
AND STRESS FRACTURES
David B. Burr and Chuck Milgrom

EXERCISE and THE STRESS RESPONSE:
THE ROLE of STRESS PROTEINS
Marius Locke and Earl G. Noble

ENDURANCE EXERCISE and ADIPOSE TISSUE
Barbara Nicklas

EXERCISE and STRESS RESPONSE
The Role of Stress Proteins

Edited by

Marius Locke & Earl G. Noble

CRC PRESS

Boca Raton   London   New York   Washington, D.C.

Library of Congress Cataloging-in-Publication Data

Exercise and stress response: the role of stress proteins / edited by Marius Locke and Earl G. Noble.
    p. cm.—(CRC series in exercise physiology)
    Includes bibliographical references and index.
    ISBN 0-8493-0458-0
    1. Heat shock proteins. 2. Stress (Physiology) 3. Exercise—Physiological aspects.
  I. Locke, Marius. II. Noble, Earl George. III. Series.

QP552.H43 E945 2002
612′.044—dc21   2002017494

Visit the CRC Press Web site at www.crcpress.com

About the Editors

Marius Locke, Ph.D., is an assistant professor in the Faculty of Physical Education and Health at the University of Toronto, Ontario. He received an honors B.A. in physical education (1984), a B.S. in biology (1987) and a Ph.D. in kinesiology (1992) from the University of Western Ontario where he also played varsity football. Dr. Locke was awarded a Natural Science and Engineering Research Council of Canada Post Doctoral Fellowship and studied at the Deborah Research Institute in Browns Mills, New Jersey from 1993–1996. In 1999, Dr. Locke received the American College of Sports Medicine's new investigator award.

Dr. Locke is a member of American College of Sports Medicine, the Canadian Society for Exercise Physiology and the Cell Stress Society International. He teaches undergraduate physical education courses on physical activity and the biophysical sciences, exercise physiology and environmental physiology as well as a graduate course on applied muscle physiology and biochemistry. Dr. Locke's research interests focus on the expression of stress proteins in striated muscle.

Earl G. Noble, Ph.D., is an associate professor in the School of Kinesiology at the University of Western Ontario, Ontario. He received his B.S. (1973) and M.S. (1975) in Kinesiology from the University of Waterloo and his Ph.D. (1980) from Washington State University.

Dr. Noble is a member of the Canadian Society for Exercise Physiology, the American College of Sports Medicine and the Cell Stress Society International and the Research Group on Biochemistry of Exercise of the International Council of Sport Science and Physical Education (UNESCO).

Dr. Noble's academic endeavors are split between teaching undergraduate and graduate students and conducting research in collaboration with the graduate students he supervises. The overriding theme of his research is to examine muscle plasticity and the manner in which muscle adapts to novel or stressful conditions, especially exercise. Over the past several years this has involved an examination of the potential role for heat shock proteins in this process.

Contributors

Marco Colavecchia
School of Kinesiology and Health
  Science
York University
Toronto, Ontario, Canada

Joseph W. Gordon
School of Kinesiology and Health
  Science
York University
Toronto, Ontario, Canada

Karyn L. Hamilton
Cardiology Research
Baylor College of Medicine
Houston, Texas

Jenny S.L. Ho
Department of Zoology
University of Toronto
Toronto, Ontario, Canada

David A. Hood
School of Kinesiology and Health
  Science
York University
Toronto, Ontario, Canada

Malcolm J. Jackson
Department of Medicine
University of Liverpool
Liverpool, United Kingdom

Yuefei Liu
Department of Sports and
  Rehabilitation Medicine
University of Ulm
Ulm, Germany

Anne McArdle
Department of Medicine
University of Liverpool
Liverpool, United Kingdom

Vandana Menon
Division of Clinical Research
New England Medical Center
Boston, Massachusetts

Pope L. Moseley
School of Medicine
University of New Mexico
Albuquerque, New Mexico

Zain Paroo
Faculty of Health Sciences, School
  of Kinesiology
University of Western Ontario
London, Ontario, Canada

Scott K. Powers
Departments of Exercise and Sport
  Sciences and Physiology
Center for Exercise Science
University of Florida
Gainesville, Florida

Arne A. Rungi
School of Kinesiology and Health
  Science
York University
Toronto, Ontario, Canada

Jeremy J. Schneider
School of Kinesiology and Health
  Science
York University
Toronto, Ontario, Canada

Joseph W. Starnes
Department of Kinesiology and
  Health Education
Cardiac Metabolism Laboratory
The University of Texas at Austin
Austin, Texas

Jürgen M. Steinacker
Department of Sports and
  Rehabilitation Medicine
University of Ulm
Ulm, Germany

Donald B. Thomason
Department of Physiology
College of Medicine
University of Tennessee Health
  Science Center
Memphis, Tennessee

J. Timothy Westwood
Department of Zoology
University of Toronto
Toronto, Ontario, Canada
and
Canadian Drosophila Microarray
  Centre
Mississauga, Ontario, Canada

Contents

chapter one

Overview of the stress response

Marius Locke

Contents

I. Introduction

In a broad sense, the word "stress" often refers to some entity that when applied to another upsets or disturbs equilibrium. These alterations or changes in stasis usually impair or alter normal function and, if left unchecked, may eventually have lethal consequences. Because almost all disciplines of study employ or describe some model capable of change, it is not surprising that stress is examined by most disciplines. From a physiological point of view, a stress can be any disruption to homeostasis. The stress-induced disruption may be at the level of the cell, tissue, organ, organ system, or whole organism. The extent of the homeostatic disruption is dependent on both the intensity and duration of the stress. At the cellular level, the investigation into the "cellular stress response" and the proteins expressed by this response, the so-called "stress proteins" (SPs) or "heat shock proteins" (HSPs), is a well-established area of research. Indeed, a PubMed computer search using the words "cells" and "stress" listed over 35,000 journal articles. Although the majority of these studies were published in journals from a wide variety of disciplines, a journal devoted exclusively to studying cell stress (and chaperones) from an interdisciplinary approach has also been established. The purpose of this chapter is to provide a brief overview of the cellular stress response. A brief description of the history of the response, from its initial discovery to the first studies involving exercise, is provided. A secondary purpose is to provide new readers with an awareness of the terms and conditions that often confuse both newcomers as well as those established in the field.

II. The nature of stress

Due to its inherent nature, the term "stress" can have different meanings to different investigators. To the exercise physiologist, the term "stress" involves perturbations to homeostasis created by exercise at, or above, the cellular level. Because exercise or physical activity is a common stress often experienced by the general populace, developing a better understanding of the consequences of the stresses it imposes may have important health implications. In addition, a more thorough understanding of the limits, as well as the mechanisms underlying these stresses, may also be relevant for enhancing performance. Despite these obvious connections between stress and exercise, a computer search using the words "exercise" and "stress proteins" listed only 66 papers. Thus, the field of investigation examining how exercise alters the "cellular stress response" has only recently emerged. Given the importance of prevention in combating disease, exercise physiologists may also have an avenue that may allow the acquisition of new information that can potentially be used to establish more appropriate exercise limits. This may allow individuals to more effectively achieve their health and fitness goals.

III. Hans Selye and the stress response

One of the first and perhaps one of the best physiological descriptions of stress came from Hans Selye, the father of stress and author of the book *The Stress of Life*. Selye defined stress as the nonspecific response of the body to any demands.[1,2] In Selye's view, stress could be either beneficial (eustress) or detrimental (distress). However, regardless of the source of stress, the physiological response was similar. Selye developed a General Adaptation Syndrome that described three stages or phases of an organism's response to stress. In the first stage, known as the alarm reaction stage, an organism

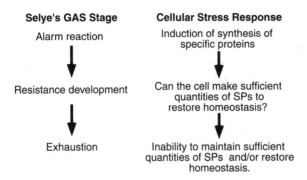

Figure 1.1   Comparison between Selye's GAS stages and induction of stress proteins.

attempts to restore homeostasis and reestablish equilibrium. If continual exposure to the stress is compatible with adaptation, the second stage — resistance development — ensues. In this stage, the signs characteristic of the alarm reaction have virtually disappeared and resistance rises above normal. However, if the organism is continuously exposed to a stress for a long period of time or if the stress is of sufficient intensity such that the adaptation capabilities are insufficient or depleted, the organism enters the exhaustion stage. In this stage, also known as the distress stage, the organism can no longer cope with the stress and, eventually, systems fail and death may ensue. Although Selye's General Adaptation Syndrome was developed to describe how organisms respond to specific stresses, it can now be used to describe how the cells of an organism respond to a myriad of stressors, including exercise. This response has become known as the cellular stress response and it fits easily within Selye's General Adaptation Syndrome (GAS) paradigm (Figure 1.1). In this scenario, after cells have been exposed to a stress, the cells respond by synthesizing SPs; this is the alarm reaction stage. If the cells can synthesize sufficient quantities of SPs to restore cellular homeostasis, resistance to the stress may be conferred and the cells survive. However, if the stress is so great in magnitude and/or duration that the quantity of SPs synthesized is insufficient to allow the cells to cope with the perturbations and stasis cannot be restored to a manageable level, the cells will enter the exhaustion stage and cell death may ensue.

IV. History of the cellular stress response

It has been known for quite some time that all nucleated cells respond to stresses capable of altering homeostasis by inducing or enhancing the synthesis of groups of highly conserved proteins.[3,4] This universal biological response has been termed the "cellular stress response" and the proteins synthesized aptly termed "stress proteins."

The first example of a cellular stress response was reported in 1962 in the form of a brief communication that showed the response of *Drosophila busckii* larvae to elevated temperatures (also termed heat shock), salicylate, and dinitrophenol.[5] These stressors were reported to cause changes in the puffing pattern of the polytene chromosomes in the salivary gland. The puffing pattern observed was reversible and dependent on environmental conditions. Because RNA synthesis was increased, these sites were suggested to be those of active genes.[5]

Perhaps due to the technology available at the time, the cellular stress response received little additional attention until 1974 when experiments using ^{35}S-methionine-labeled proteins separated by SDS-PAGE clearly showed the synthesis of new proteins during/following exposure to elevated temperatures, or "heat shock."[6] Because heat shock was one of the first stressors applied, the newly synthesized proteins were termed "heat shock proteins" (HSPs). One of the first reports of HSPs in vertebrates showed that in addition to heat, other stressors also induced the synthesis of HSPs.[7] As the list of stressors capable of causing the "heat shock response" and inducing the HSPs grew, the terms "cellular stress response" and "stress protein" were introduced to more accurately reflect the myriad of stressors capable of inducing the response. Today, the terms HSPs and SPs are often used interchangeably although there are differences.

As more studies were conducted, it became apparent that not all stressors induced the same SPs.[8–12] That is, cells responded to different stressors by synthesizing different proteins. For example, the proteins induced following heat stress were different in molecular weight than the proteins induced in cells deprived of glucose.[8,9,11,12] Thus, at least two distinct sets of proteins were induced in cells. The most prominent group, the HSPs, was induced by heat shock and other protein-damaging stressors, while a second group was induced following perturbations to glucose homeostasis (glucose and calcium impairment, as well as agents that impair protein glycosylation). Due to their induction following glucose perturbations, the latter group became known as glucose regulated proteins (GRPs). GRPs are generally not induced in cells by heat shock.

A distinguishing feature between the HSPs and the GRPs is the presence of a specific regulatory sequence in the promotor region of all HSP genes. This sequence is known as the heat shock element (HSE) and confers heat inducibility.[13,14] The HSE is the binding site for regulatory proteins known as heat shock transcription factors (HSFs).[15,16] HSFs mediate the expression of the HSP genes during conditions of stress as well as during some nonstressful conditions.[17] At present, a number of HSFs have been identified, some of which exhibit isoforms generated by alternative splicing.[17,18] Because GRPs do not possess an HSE, they are not mediated by the HSFs and are not heat inducible. Thus, although HSPs and GRPs can both be considered SPs, each group is a separate set of proteins capable of being selectively induced.[8–12,19]

During the 1980s, recombinant DNA techniques became popular and some of the HSPs were among the first genes to be cloned. The heat shock response and induction of HSPs became a convenient model system for studying inducible genes, and an increased awareness of HSPs (and SPs) was established primarily by a small number of basic researchers. As more SPs were identified, and as more researchers examined their model systems for SPs, the scientific prominence of SPs gradually increased. Today, SPs are being studied by many disciplines and play important roles in areas ranging from immunology to myocardial protection. In addition, SPs are being studied for their roles in protein transport, protein synthesis, as well as their ability to protect cells and tissues from reactive oxygen species. These topics are addressed in detail in subsequent chapters.

Given that the stressors known to induce SPs in cells *in vitro* were very similar to the homeostatic perturbations known to occur in cells and tissues during exercise, it was not surprising that exercise physiologists became interested in examining SPs during and following exercise. The often-observed exercise-induced hyperthermia, combined with other cellular perturbations, implied that exercise might be capable of inducing SPs. The first study to investigate the expression of SPs during exercise involved mice that were exercised by swimming.[20] Although no increase in SP content was observed, it set the stage for additional exercise studies,[21-27] and it is now generally accepted that certain modes and intensities of exercise are capable of inducing SPs in mammals, including humans. However, the mechanism(s) responsible for the exercise induction of SPs and the exact significance of the induction of SPs by exercise are less well understood and a current area of investigation.

V. Stress protein inducers

Because temperatures of 43 to 45°C were often used to induce HSPs in mammalian cell lines, SPs were initially viewed as an *in vitro* phenomenon that occurred primarily when cells were subjected to supraphysiological levels of stress. Thus, the physiological role(s) of SPs in cells was unclear and the exact function(s) of the various SPs remained largely unknown. Eventually, SPs were shown to be expressed in cells in response to physiologically relevant stressors.[21-27] SPs were also shown to be expressed in cells during unstressed conditions and in some cases in a tissue-specific manner.[22,28-30] Stressors capable of inducing SPs can be classified as either environmental stresses, pathophysiological stresses, or as nonstress conditions.[17] Environmental stressors are known to include, but are not limited to, temperature changes, exposure to transition metals, oxygen-derived free radicals, amino acid analogues, and inhibitors of energy metabolism. Some of the pathophysiological stresses may include fever, ischemia, oxidant injury, aging, hypertrophy, as well as viral and bacterial infections. Nonstressful conditions include growth factors, specific stages of the cell cycle, as well as processes that occur during cell development and differentiation. While these classifications are useful, it is now generally accepted that any stress capable of damaging or denaturing cellular proteins will likely induce an SP response.

VI. Classification of stress proteins

All proteins exhibit a charge and molecular weight; hence, these characteristics are used to separate and identify SPs. Although there are numerous
methods for determining these characteristics, a common method is electrophoretic separation of proteins by polyacrylamide gels. Thus, the number
following the HSP or GRP abbreviation refers to its approximate molecular
weight in kilodaltons (kDa). For example, HSP60 refers to a heat shock
protein with a molecular weight of 60,000 daltons or 60 kilodaltons (60 kDa).
As mentioned previously, although many stresses are capable of inducing
these proteins, many SPs are still referred to by their original name of heat
shock protein because heat stress/shock was the initial stressor used to
characterize these proteins. Because many SPs express isoforms, there can
also be confusion as to the exact SP being examined. For example, HSP70 is
often used to refer to both the entire family of proteins with an approximate
molecular weight of 70 kDa, as well as a specific member (the inducible member
also known as HSP72) of this family. Additional confusion can arise from
the same SP exhibiting a slightly different molecular weight in different
species. For example, HSP27 and HSP25 are both used to refer to the same
protein in human and rodent species, respectively.[32] Thus, there can be uncer-
tainty about the exact SP under investigation.

SPs are often placed into groups or families based on their relative molecular weight and DNA sequence. The exact number of these groups varies but
usually there are at least three main groups. Most groupings generally identify
a group of low-molecular-weight SPs, the HSP70 family of SPs, and a group
of high-molecular-weight SPs. The high-molecular-weight SPs may or may
not include SPs with an approximate molecular weight of 90 kDa. The 90-kDa
family of SPs is often placed in a separate group and the high-molecular-
weight SPs used to refer to SPs of approximately 100 to 110 kDa.

Most group classifications of SPs also refer to a low-molecular-weight
group of SPs. While this term is generally used to refer to a family of isoforms
with a molecular weight of approximately 23 to 27 kDa, it should be noted
that there are many other low-molecular-weight proteins that are also considered SPs but are not members of this family. For example, heme oxygenase[33]
and ubiquitin[34] are both proteins that have been identified as SPs with a low
molecular weight but may have little similarity to the proteins traditionally
identified as the low-molecular-weight SPs. Thus, unless referring to a specific family of SPs (i.e., HSP70 or HSP90), the general groupings of SPs are
of limited value.

VII. Constitutive expression of stress proteins

Although the term "stress protein" was originally intended to describe proteins that are induced during or following stress, it has become apparent
that many SPs are constitutively expressed. That is, they are expressed in
cells during "unstressed" conditions.[22,28,29,31] For example, the 70-kDa family

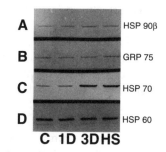

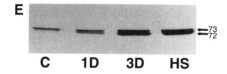

Figure 1.2 Myocardial stress protein content after exercise or heat shock. Rats were subjected to: no stress [C], one bout of treadmill running (30 m/min for 60 min) [1D], three bouts of treadmill running on three consecutive days (30 m/min for 60 min) [3D] or a 15-minute, 42°C heat shock [HS]. All animals were allowed to recover for 24 hours and the myocardial SP content assessed by SDS-PAGE followed by Western blotting. Panel A: HSP90β. Panel B: GRP75. Panel C: HSP70 (both HSP72 and HSC73). Panel D: HSP60. Panel E: an enlargement of a blot similar to that shown in panel C. Both HSP72 and HSC73 are indicated by arrows.

of stress proteins is known to have both an *inducible* isoform (HSP72) and a *cognate* isoform (HSC73).[21,22] The former, as its name suggests, was initially observed to be absent or at low levels in most cells during unstressed conditions, but increased following exposure to stresses.[21,35] In contrast, the cognate isoform is readily expressed in most cells during unstressed conditions and generally does not demonstrate high levels of induction following exposure to stresses.[21,35] Figure 1.2 shows portions of Western blots demonstrating the myocardial content of four SPs, 24 hours after rats were subjected to either one exercise bout (treadmill running at 30 meters/minute for 60 min), three exercise bouts (on three consecutive days), or a 15-minute 42°C heat shock. Figure 1.2 panel C shows the typical increased expression of HSP70 detected in the rat myocardium 24 hours after either three bouts of exercise (on three consecutive days) or a 15-minute, 42°C heat shock. Panel E, a magnification of a blot similar to that in panel C, shows both a faster migrating inducible HSP70 isoform (HSP72) and a slightly slower migrating (less inducible) cognate isoform (HSC73). These isoforms are more easily resolved and identified using two-dimensional gel electrophoresis or HSP70 isoform-specific antibodies, respectively. From these blots it can also be determined that detectable quantities of certain SPs, both HSPs and GRPs, are detectable in hearts from unstressed controls. Furthermore, the myocardial content of certain SPs (HSP90β, GRP75, HSP60) is not increased after either exercise or heat shock. This demonstrates the differential patterns of SP induction that are often observed.

Because the inducible isoform of the HSP70 family (i.e., HSP72) can now be readily detected in many cells and tissues during unstressed conditions, it is not uncommon to refer to the constitutive expression of the inducible isoform. Indeed, certain cells and tissues, including slow skeletal muscles, have been shown to express high levels of the inducible isoform, HSP72, such that its level in these cells is comparable to the HSP72 levels in cells subjected to stress.[22,29,30] The exact significance of these findings remains unclear, but it serves to emphasize that the term "stress protein" may be somewhat of a misnomer because many of the so-called stress proteins are also detectable during unstressed conditions in certain cells or tissues. The constitutive expression of SPs also suggests that using SPs as markers of stress may also require additional examination.

VIII. Stress protein function

SPs are highly conserved between species and have been observed in every cell type examined to date, suggesting that SPs may have a universal and thus important biological function(s). Although it was recognized that SPs were the cell's response to stress, their exact functions have remained more elusive. Because HSPs were first induced by heat shock, it was logical to examine their role in protecting cells during exposure to increases in temperature. Not surprisingly, several early studies examined the relationship between HSP expression and thermotolerance.[36,37] The expression of certain HSPs was shown to correlate with the acquisition and decay of thermotolerance. This was shown using manipulations at both the protein and DNA levels.[37,38]

Although possible roles for the various SPs were suggested, it was not until certain mammalian 70-kDa heat shock protein isoforms were shown to be structurally and functionally related to proteins involved in releasing clathrin triskelions from coated vesicles that insight into the function of certain SPs was provided.[39,40] The same SPs (HSP70 members) were also shown to be involved in facilitating the translocation of polypeptides into microsomes and mitochondria.[41,42] Thus, SPs were shown to play important roles in maintaining and facilitating the correct protein folding, assembly, and transport of protein structures within cells.[43] To accurately describe these functions, the term "molecular chaperones" was developed and thus some SPs are correctly referred to as molecular chaperones.[44,45] SPs have since been shown to be involved in many protein-related or chaperone-like functions, including protein synthesis, degradation, trafficking, transport, and repair.[46,47] Given their intimate relationship with proteins and protein-related functions within the cell, as well as the importance of these functions to the cell during times of stress, it is not surprising that SPs have been shown to play important roles in cell protection. Indeed, the ability to protect cells during episodes of stress is thought to be due to the chaperone ability of SPs.

If the induction of HSPs by heat stress protected cells from subsequent heat stresses, then a logical extension was to determine if HSPs induced by heat could also protect cells or tissues from a second stressor. This concept, known as cross-tolerance, occurs when stressor A, which in most cases is heat, is used to elevate HSPs to subsequently provide protection against stressor B. Cross-tolerance has important implications for exercise and it has been examined using a rat model.[26,48-52] HSPs were induced in the rat heart by raising core body temperature to 42°C for 15 minutes, after which the rats were allowed to recover for 24 hours to increase myocardial HSP content. Hearts were subsequently removed and placed on an isolated heart apparatus (Langendorff) where an ischemic stress (30 minutes) was applied. The prior induction of HSPs in the heart minimized the myocardial damage from ischemia, such that hearts from heat-shocked animals demonstrated an enhanced recovery of myocardial contractile force and a reduction in creatine kinase release during a subsequent reperfusion period.[48-52] This protective ability conferred by HSPs has since been confirmed using transgenic animals capable of overexpressing certain HSPs.[53-55] The HSP mediated protection has also been examined in skeletal muscle.[56] From an exercise viewpoint, the concept of cross-tolerance provides a possible mechanism whereby exercise, through induction of SPs, may confer protection to cells and tissues. Preliminary studies using exercise as a stressor have shown a similar cross-tolerance effect to the heart.[26] As more exercise studies are conducted, greater insight into the role that SPs play in conferring protection to cells and tissues will be ascertained.

IX. Conclusions

Our knowledge regarding the cellular stress response, as well as the proteins involved in the response, continues to increase and remains an active area of investigation. Although exercise has long been considered a stress, it has only recently been examined as a natural physiological method for inducing the protective cellular stress response and only a few studies have examined the relationship between exercise and the stress response. It appears that exercise is capable of altering cellular homeostasis to such an extent that the cellular stress response is activated and SPs accumulate in certain tissues. However, the exact significance of the increased SP content, as well as the mechanism(s) by which exercise induces SPs and confers protection to cells and tissues, remain to be determined. This will likely be an important area of future investigation. The possibility of using a nonpharmacological method, such as exercise, as a means to activate or harness a cell's endogenous system of protection is an exciting possibility (SPs) that may have important health implications. The subsequent chapters in this book provide greater insight and detail into the relationship between cellular stress response and topics relevant to exercise.

Acknowledgments

This work was supported in part by grants from the National Sciences and Engineering and Research Council of Canada and the Heart and Stroke Foundation of Canada.

References

1. Selye, H., *The Stress of Life*, 2nd ed., McGraw-Hill, New York, 1976.
2. Perdrizet, G. A., Hans Selye and beyond: responses to stress, *Cell Stress Chaperones*, 2, 214, 1997.
3. Subjeck, J. R. and Shyy, T.-T., Stress protein systems of mammalian cells, *Am. J. Physiol.*, 250, C1, 1986.
4. Morimoto, R. I., Tissières, A., and Georgopoulos, C., Eds., *Stress Proteins in Biology and Medicine*, Cold Spring Harbor Laboratory Press, Cold Spring Harbor, New York, 1990.
5. Ritossa, F., A new puffing pattern induced by temperature shock and DNP in *Drosophila*, *Experientia*, 15, 571, 1962.
6. Tissières, A., Mitchell, H. K., and Tracy, U. M., Protein synthesis in salivary glands of *Drosophila melanogaster*: relation to chromosome puffs, *J. Mol. Biol.*, 84, 389, 1974.
7. Kelley, P. M. and Schlesinger, M. J., The effect of amino acid analogues and heat-shock on gene expression in chicken embryo fibroblasts, *Cell*, 15, 1277, 1978.
8. Welch, W. J. et al., Biochemical characterization of the mammalian stress proteins and identification of two stress proteins as glucose and Ca^{2+}-ionophore-regulated proteins, *J. Biol. Chem.*, 258, 7102, 1983.
9. Lee, A. S., The accumulation of three specific proteins related to glucose-regulated protein in a temperature-sensitive hamster mutant cell line K12, *J. Cell. Physiol.*, 106, 119, 1981.
10. Hightower, L. E. and White, F. P., Cellular response to stress; comparison of a family of 71–73 kilodalton proteins rapidly synthesized in rat tissue slices and canavanine-treated cells in culture, *J. Cell. Physiol.*, 108, 261, 1981.
11. Whelan, S. A. and Hightower, L. E., Differential induction of glucose-regulated and heat shock proteins: effects of pH and sulfhydryl-reducing agents on chicken embryo cells, *J. Cell. Physiol.*, 125, 251, 1985.
12. Sciandra, J. J. and Subjeck, J. R., The effects of glucose on protein synthesis and thermosensitivity in Chinese hamster ovary cells, *J. Biol. Chem.*, 258, 12091, 1983.
13. Pelham, H. R. B., A regulatory upstream promotor element in *Drosophila* HSP70 heat shock gene, *Cell*, 30, 517, 1982.
14. Amin, J., Ananthan, J., and Voellmy, R., Key features of heat shock regulatory elements, *Mol. Cell. Biol.*, 8, 3761, 1988.
15. Westwood, J. T., Clos, J., and Wu, C., Stress-induced oligomerization and chromosomal relocalization of heat-shock factor, *Nature*, 353, 822, 1991.
16. Sarge, K. D. et al., Cloning and characterization of two mouse heat shock factors with distinct inducible and constitutive DNA-binding ability, *Genes Dev.*, 5, 1902, 1991.
17. Morimoto, R. I., Sarge, K. D., and Abravaya, K., Transcriptional regulation of heat shock genes, *J. Biol. Chem.*, 267, 21987, 1992.

18. Goodson, M. L. and Sarge, K. D., Regulated expression of heat shock factor 1 isoforms with distinct leucine zipper arrays via tissue-dependent alternative splicing, *Biochem. Biophys. Res. Commun.*, 211, 943, 1995.

19. Mizzen, L. A. et al., Identification, characterization, and purification of two mammalian stress proteins present in mitochondria, GRP75, a member of the HSP70 family and HSP58, a homolog of the bacterial groEL protein, *J. Biol. Chem.*, 264, 20664, 1989.

20. Hammond, G. L., Lai, Y.-K., and Markert, C. L., Diverse forms of stress lead to new patterns of gene expression through a common and essential metabolic pathway, *Proc. Natl. Acad. Sci.*, 79, 3485, 1982.

21. Locke, M., Noble, E. G., and Atkinson, B. G., Exercising mammals synthesize stress proteins, *Am. J. Physiol.*, 258, C723, 1990.

22. Locke, M., Noble, E. G., and Atkinson, B. G., Inducible isoform of HSP70 is constitutively expressed in a muscle fiber type specific pattern, *Am. J. Physiol.*, 261, C774, 1991.

23. Salo, D. C., Donovan, C. M., and Davies, K. J. A., HSP70 and other possible heat shock or oxidative stress proteins are induced in skeletal muscle, heart, and liver during exercise, *Free Radic. Biol. Med.*, 11, 239, 1991.

24. Thompson, H. S. and Scordilis, S. P., Ubiquitin changes in human biceps muscle following exercise-induced damage, *Biochem. Biophys. Res. Commun.*, 204, 1193, 1994.

25. Skidmore, R. et al., HSP70 induction during exercise and heat stress in rats: role of internal temperature, *Am. J. Physiol.*, 268, R92, 1995.

26. Locke, M. et al., Enhanced post-ischemic recovery following exercise induction of HSP70 in rat heart, *Am. J. Physiol.*, 269, H320, 1995.

27. Locke, M., Ianuzzo, S. E., and Ianuzzo, C. D., Activation of the heat shock transcription factor in the rat heart following heat shock and exercise, *Am. J. Physiol.*, 268, C1387, 1995.

28. Locke, M., Tanguay, R. M., and Ianuzzo, C. D., Constitutive expression of HSP 72 in swine heart, *J. Mol. Cell. Cardiol.*, 28, 467, 1996.

29. Locke, M. and Tanguay, R. M., Increased HSF activation in muscles with a high constitutive HSP70 expression, *Cell Stress Chaperones*, 1, 189, 1996.

30. Locke, M., Heat shock factor activation and HSP72 accumulation in aged skeletal muscle, *Cell Stress Chaperones*, 5, 45, 2000.

31. Locke, M. et al., Shifts in the proportion of type I fibers in rat hindlimb muscle are accompanied by changes in HSP72 content., *Am. J. Physiol.*, 266, C1240, 1994.

32. Arrigo, A.-P. and Landry, J., Expression and function of the low-molecular-weight heat shock proteins, in *The Biology of Heat Shock Proteins and Molecular Chaperones*, Morimoto, R. I., Tissières, A., and Georgopoulos, C., Eds., Cold Spring Harbor Laboratory Press, Cold Spring Harbor, New York, 1994, 335.

33. Shibahara, S., Muller, R. M., and Taguchi, H., Transcriptional control of rat heme oxygenase by heat shock, *J. Biol. Chem.*, 262, 12889, 1987.

34. Bond, U. and Schlesinger, M. J., The chicken ubiquitin gene contains a heat shock promotor and expresses an unstable mRNA in heat-shocked cells, *Mol. Cell. Biol.*, 6, 4602, 1986.

35. Locke, M. and Atance, J., The myocardial heat shock response following sodium salicylate treatment, *Cell Stress Chaperones*, 5, 359, 2000.

36. Landry, J. et al., Thermotolerance and heat shock proteins induced by hyperthermia in rat liver cells, *Int. J. Radiat. Oncol. Biol. Phys.*, 8, 59, 1982.

37. Riabowol, K. T., Mizzen, L. A., and Welch, W. J., Heat shock is lethal to fibroblasts microinjected with antibodies against HSP70, *Science*, 242, 433, 1988.

38. Johnston, R. N. and Kucey, B. L., Competitive inhibition of HSP70 gene expression causes thermosensitivity, *Science*, 242, 1551, 1988.

39. Ungewickell, E., The 70-kd mammalian heat shock proteins are structurally and functionally related to the uncoating protein that releases clathrin triskelia from coated vesicles, *EMBO*, 4, 3385, 1985.

40. Chappell, T. et al., Uncoating ATPase is a member of the 70 kilodalton family of stress proteins, *Cell*, 45, 3, 1986.

41. Deshaies, R. J. et al., A subfamily of stress proteins facilitates translocation of secretory and mitochondrial precursor polypeptides, *Nature*, 332, 800, 1988.

42. Chirico, W. J., Waters, M. G., and Blobel, G., 70K heat shock related proteins stimulate protein translocation into microsomes, *Nature*, 332, 805, 1988.

43. Beckmann, R. P., Mizzen, L. A., and Welch, W. J., Interaction of HSP70 with newly synthesized proteins: implications for protein folding and assembly, *Science*, 248, 850, 1990.

44. Ellis, J., Proteins as molecular chaperones, *Nature*, 328, 378, 1987.

45. Ellis, R. J. and van der Vies, S. M., Molecular chaperones, *Annu. Rev. Biochem.*, 60, 321, 1991.

46. Chiang, H.-L. et al., A role for a 70-kilodalton heat shock protein in lysosomal degradation of intracellular proteins, *Science*, 246, 382, 1989.

47. Baler, R., Welch, W. J., and Voellmy, R., Heat shock gene regulation by nascent polypeptides and denatured proteins: HSP70 as a potential autoregulatory factor, *J. Cell. Biol.*, 117, 1141, 1992.

48. Currie, R. W. et al., Heat-shock response is associated with enhanced postischemic ventricular recovery, *Circ. Res.*, 63, 543, 1988.

49. Currie, R. W. and Karmazyn, M., Improved post-ischemic ventricular recovery in the absence of changes in energy metabolism in working rat hearts following heat shock, *J. Mol. Cell. Cardiol.*, 22, 631, 1990.

50. Karmazyn, M., Mailer, K., and Currie, W. R., Acquisition and decay of heat-shock-enhanced postischemic ventricular recovery, *Am. J. Physiol.*, 259, H424, 1990.

51. Currie, R. W. and Tanguay, R. M., Analysis of rat heart RNA for transcripts for catalase and SP71 after *in vivo* hyperthermia, *Biochem. Cell. Biol.*, 67, 375, 1991.

52. Currie, R. W., Tanguay, R. M., and Kingma, J. G., Jr., Heat-shock response and limitation of tissue necrosis during occlusion/reperfusion in rabbit hearts, *Circulation*, 87, 963, 1993.

53. Plumier, J.-C. L. et al., Transgenic mice expressing the human heat shock protein 70 have improved post-ischemic myocardial recovery, *J. Clin. Invest.*, 95, 1854, 1995.

54. Marber, M. S. et al., Overexpression of the rat inducible 70-kD heat stress protein in a transgenic mouse increases the resistance of the heart to ischemic injury, *J. Clin. Invest.*, 95, 1446, 1995.

55. Radford, N. B. et al., Cardioprotective effects of 70-kDa heat shock protein in transgenic mice, *Proc. Natl. Acad. Sci., U.S.A.*, 93, 2339, 1996.

56. Garramone, R. R. et al., Reduction of skeletal muscle injury through stress conditioning using the heat shock response, *Plast. Reconstr. Surg.*, 93, 1242, 1994.

chapter two

Transcriptional regulation of the mammalian heat shock genes

Jenny S. L. Ho and J. Timothy Westwood

Contents

I. Introduction

The ability of cells to respond rapidly and appropriately to external and physiological stresses is essential for their survival in an ever-changing environment. There are many forms of stress that result in the transcriptional induction of specific genes, and the products of those genes generally help the organism cope with that damage caused by the stress. Such stresses include oxidative/reactive oxygen stress,[1-4] toxic heavy metals,[5-7] and heat shock/proteotoxic stress. This chapter focuses on the regulation of one of these stress responses: the heat shock response.

The heat shock response occurs in organisms as diverse as bacteria and humans and is induced by exposure to elevated temperatures as well as a variety of chemical and physiological stresses. The response involves the transcriptional induction of the heat shock genes, which encode a set of conserved polypeptides known as the heat shock proteins (HSPs). These HSPs function as molecular chaperones that assist in protein folding, translocation, and degradation.[8-13] Because of their capacity to repair and prevent protein damage and aggregation, the HSPs confer stress tolerance to the cells against a variety of proteotoxic agents that induce abnormal proteins and protein damage.[12,14-16] Some of these proteotoxic agents include heat, amino acid analogues, transition series metals, and alcohols.[17-20] In addition to external heat and the various chemical inducers, the expression of HSPs is invoked by a number of physiological stresses, such as exercise, which increases core body temperatures,[21] and pathophysiological stresses and disease states such as fever, inflammation,[22] infection,[23] and ischemia.[24] The precise functions of the HSPs under these conditions, however, are not well defined. In essence, the heat shock response plays a pivotal role in the maintenance of protein homeostasis, cellular survival, and recovery during environmental and physiological stresses; as a consequence, intensive efforts and progress have been made toward the elucidation of the processes that are involved in the regulation and execution of the defense response.

The heat shock response was first discovered in 1962 by Ritossa,[25] who described the induction of a set of chromosomal puffs on the polytene chromosomes of *Drosophila busckii* upon exposure to heat, dinitrophenol, or sodium salicylate. Subsequently, Tissières and co-workers[26] observed that the stress-induced puffs coincided with the synthesis of a set of new proteins, which were later isolated and characterized as the products of the heat shock genes.[27] The HSPs can be grouped into several families: HSP100, HSP90, HSP70, HSP60, HSP40, and the small HSPs. As well, the HSP families contain constitutively expressed members known as the heat shock cognates, or hscs, which function as molecular chaperones to facilitate the folding and assembly of proteins under normal conditions.[12] In addition to the changes in gene transcription, the heat shock response affects other molecular processes. For example, preferential translation of heat shock gene transcripts during heat stress occurs in a number of organisms (reviewed in References 27 and 28).

In eukaryotes, the expression of the inducible heat shock genes is primarily regulated at the level of transcription[27,28] by the heat shock transcription factor (HSF), whose activity is in turn tightly controlled by a complex, multiple-step cascade of events that is induced upon exposure to stress. As a consequence, the elucidation of the mechanisms involved in the regulation of HSF activity is central to the understanding of the heat shock response as a whole.

II. Heat shock transcription factors

This chapter mainly focuses on animal, and in particular mammalian, heat shock transcription factors (HSFs) and their regulation. Several excellent reviews of HSFs have been published in recent years[30–34] and more information on HSFs, including those found in yeasts and plants, can be found in these reviews. The cloning of HSF genes from various organisms has led to the finding of multiple HSF proteins. Whereas the fruitfly (*Drosophila melanogaster*)[35] and yeast (*Saccharomyces cerevisiae*)[36] possess a single HSF, two HSF genes have been isolated from mouse (HSF1 and HSF2)[37] and three each from human (HSF1, HSF2, HSF4)[38,39] and chicken (HSF1, HSF2, HSF3).[40] A phylogenetic comparison at the amino acid level of all the sequenced animal HSF genes is shown in Figure 2.1. Vertebrate HSF1s and HSF2s cluster together well and divergence within clusters is in good agreement with expected evolutionary divergence. As an example, the similarity between the mouse and human HSF1 and the mouse and human HSF2 is 90% and 95%, respectively.[41] It should be noted that certain protein domains of the HSF molecule, in particular the DNA binding and the oligomerization (heptad repeat) domains (see Figure 2.2A and Section III), are very highly conserved among all of the HSF molecules. Thus, in vertebrates in which there are two or more HSF genes, the homology of HSF types (e.g., HsHSF1 vs. MmHSF1) is much higher than it is to other HSFs of a different type within a species (e.g., HsHSF1 vs. HsHSF2). This supports the notion that for those species containing multiple HSFs, each HSF likely has a specific function(s) as opposed to having redundant functions.

HSF1 is known to be the principal stress-responsive transcription factor[37,42] and is the most extensively studied HSF homologue thus far. In contrast, much less is known about HSF2 and HSF4. Although it is insensitive to typical stress stimuli, studies have implicated a role of HSF2 in certain developmental processes such as spermatogenesis,[43] mouse early embryonic development,[44,45] and hemin-induced erythroid differentiation in human K562 cells.[46,47] While the expression of HSF1 and HSF2 appears to occur in all cells, HSF4 expression seems to be specific to the heart, skeletal muscle, and brain.[39] Transcripts for mouse HSF4 can be alternatively spliced to yield HSF proteins with different activities; one of these appears to lack transcriptional activity and may act as a repressor of heat shock genes while the other may act as a transcriptional activator.[39,48] Alternatively spliced transcripts for mouse HSF1 and HSF2 have also been reported and in the case of HSF2 the two isoforms appear to have different transcriptional activities *in vitro*.[49,51] The relative proportions of the polypeptides derived from the alternatively spliced HSF transcripts, as well as their function inside cells, remain an area that requires more investigation.

As mentioned, HSF1 is the main HSF homologue that is responsive to stresses caused by heat and various chemical inducers. Under normal conditions, the constitutively expressed HSF1 is maintained as a latent monomer, which, upon heat shock, associates with two other HSF1 molecules to

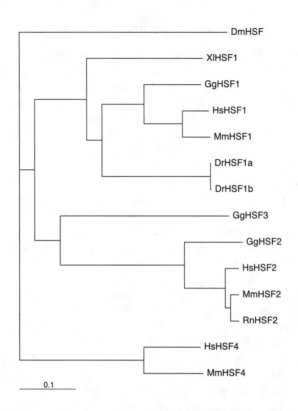

Figure 2.1 Phylogenetic relationship of the animal heat shock factor proteins. A multiple alignment was performed on the amino acid sequences of all of the animal HSF proteins found in GenBank using CLUSTAL_X version 1.8 (http://www-igbmc. u-strasbg.fr/BioInfo/ClustalX/Top.html)[162] and an unrooted tree was constructed using TreeView 1.6.5.[163] The scale bar (0.1) is substitutions per site; that is, a branch of that length has, on average, 0.1 amino acid substitutions per site. Abbreviations and accession numbers (acc. nos.): HSF, heat shock factor; Dm, *Drosophila melanogaster* (fruit fly), acc. no. M60070.1; Xl, *Xenopus laevis* (South African clawed frog), acc. no. JC4199; Gg, *Gallus gallus* (chicken), acc. nos., GgHSF1, P38529, GgHSF2, P38530, GgHSF3, P38531; Hs, *Homo sapiens* (human), acc. nos., HsHSF1, NM_005526.1, HsHSF2, M65217.1, HsHSF4, NM_001538.1; Mm, *Mus musculus* (mouse), acc. nos., MmHSF1, P38532, MmHSF2, P38533, MmHSF4, Q9R0L1; Dr, *Danio rerio* (zebrafish), acc. nos., DrHSF1a, AAF72750, DrHSF2a, AAF72751; Rn, *Rattus norvegicus* (rat), acc. no. NP_113882.

form a homotrimer.[41,52–56] *Saccharomyces cerevisiae* HSF monomers have the ability to bind heat shock gene promoter elements but metazoan monomers do not. The trimerization of HSF1 dramatically increases its DNA-binding affinity to a conserved *cis*-acting element, called the heat shock element (HSE), found in multiple copies at the promoters of the heat shock genes.[57] The heat shock element consists of three contiguous inverted repeats of a pentameric unit with the consensus sequence 5′ nGAAn 3′, where each subunit of the HSF1 trimer interacts with one of the three repeats.[52,58–60] In many heat shock gene promoters, the first nucleotide in the pentameric sequence is an adenosine.[60] Moreover, the binding of HSF1 trimers to adjacent HSEs is cooperative, further increasing its binding affinity to the heat shock gene promoters.[60-63]

The oliogomerization states for HSF2 and HSF4 appear to vary somewhat from HSF1. For example, under normal and heat shock conditions, HSF2 is dimeric and lacks HSE binding ability.[47] Upon exposure to hemin, HSF2 trimerizes and binds to HSEs.[47] In addition, the optimal HSE sequence for HSF2 binding is slightly different from that for HSF1.[62] HSF4 appears to be missing protein domains that suppress oligomerization and therefore is constitutively trimeric and HSE-bound.[39,48]

Deletion of the sole HSF gene in *Drosophila* is lethal.[64] Using a temperature-sensitive mutant allele of the HSF gene, Jedlicka and co-workers[64] found that HSF protein was dispensable for general cell growth and viability in *Drosophila* but it was required for oogenesis, early larval development, and for the ability to survive extreme heat stress. In mice, disruption of the HSF1 gene eliminates the heat shock response.[65] In addition, acquired thermotolerance is prevented and the susceptibility of cells to heat-induced apoptosis is increased in cell lines derived from HSF1$^{-/-}$ mice.[65] While HSF1$^{-/-}$ mice survive to adulthood, they exhibit a number of detrimental phenotypes, including female infertility, placental defects, and postnatal growth retardation.[66] HSF1 does not appear to be required for oogenesis in mice as oocytes were ovulated and fertilized in HSF1$^{-/-}$ females but embryos from these females mostly arrested at the 1-2 cell stage.[67]

III. Functional organization of HSF1

The activation of HSF1 into a fully competent transcription factor is a multi-faceted process, and this is reflected in the various structural domains found in HSF1. The human and mouse HSF1s are 529 and 503 amino acids in length, respectively. Found near the amino (N)-terminus of the protein are two domains that are extensively conserved in evolution: the DNA-binding domain and the trimerization domain, separated by a short flexible linker region (see Figure 2.2A). The crystal structure of the *Kluyveromyces lactis* HSF DNA-binding domain[68] and the solution structure of the *Drosophila* HSF DNA binding domain[69,70] reveal a helix-turn-helix DNA-binding motif. Carboxy (C)-terminal to this DNA-binding domain is the trimerization domain that consists of three arrays of hydrophobic heptad repeats, where the first and fourth positions within each repeat are composed of hydrophobic residues (Figure 2.2A). The organization of the heptad repeats allows the assembly of HSF monomers into homotrimers via the formation of a triple-stranded, α-helical coiled-coil structure at the trimerization domain.[71] Near the C-terminal end of all HSF1s is a third region of conservation composed of an array of hydrophobic heptad repeats that function to suppress trimerization (Figure 2.2A and B).[72,73] The acidic transactivation domain maps to the C-terminal part of all HSFs and its activity is negatively regulated by a central regulatory

A.   **Human HSF1**

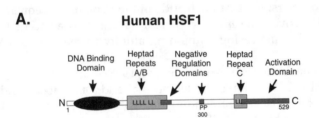

B.   **HSF1 Activation/Attenuation**

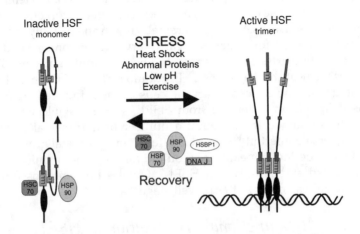

Figure 2.2   Functional organization and activation of human HSF1. A: Functional organization of HSF1. Functional protein domains are identified and their positions on the HSF1 molecule are indicated. N and C are the amino and carboxyl termini of HSF1. Numbers refer to amino acid position and P is a phosphorylation site. B: Model of the activation and deactivation of HSF1. See text for details.

domain located between the DNA-binding domain and the transactivation domain.[74-77] Finally, a nuclear localization signal for human HSF1 has been placed near the N-terminal region preceding the first hydrophobic heptad repeat of the trimerization domain.[76]

IV. Regulation of HSF1 activity

The activity of HSF1 appears to be regulated at a number of steps.[31,41,78,79] In fact, the regulation of HSF1 activity can be viewed as a concerted series of events that is dependent on the intrinsic structural and conformational characteristics of the transcription factor, as well as its interaction with various cellular regulators to bring about a finely tuned, tightly regulated response that begins with the detection and transduction of the stress signal, leading to the activation of the transcription factor and ultimately to the transcriptional activation of the heat shock genes. A number of events have been proposed to be involved in HSF1 regulation, including (1) the trimerization of HSF1 and the acquisition of DNA-binding ability,[41,53–56] (2) the achievement of trans-activation competence,[31] and (3) the attenuation of HSF1 activity.[80] In addition, cellular regulators such as the HSPs,[81–86] as well as phosphorylation activity,[87–89] may participate in the regulation of one or more of the above events.

Certain authors have suggested that an additional step, one that precedes the three steps listed above, is involved in regulating HSF activity. This step is the stress-induced translocation of HSF1 from the cytoplasm into the nucleus.[32,41,55] A number of recent studies have cast doubt on whether unstressed HSF1 is a cytoplasmically localized protein and these studies are discussed below.

A.   Cellular localization of HSF1

While all studies to date indicate that the active trimeric HSF1 is located in the nucleus, the cellular localization of the latent monomeric form of HSF1 remains a matter of debate. In some biochemical fractionation studies done in human K562 cells, HeLa cells, and mouse 3T3 cells,[41,47,55] HSF1 is found in the cytoplasmic fraction of unstressed cells. Upon heat treatment, HSF1 associates predominantly with the nuclear fraction.[41,47,55] In one immunocytochemical study with 3T3 and HeLa cells, unstressed HSF1 is distributed in a diffuse pattern in both the nucleus and the cytoplasm, while heat-stressed HSF1 is found exclusively in the nucleus.[41] From these studies, it would appear that the heat-induced activation of HSF1 might include a cytoplasmic-to-nuclear translocation event. The mapping of a nuclear localization signal at a region overlapping the DNA-binding domain and the trimerization domain leads to the proposal that heat-induced changes in HSF1 conformation or interaction with cellular regulatory proteins may unmask the nuclear localization signal, thus promoting the translocation of HSF1 upon heat shock.[76]

In contrast, other immunocytochemical studies done in human and mouse cells in culture,[79,90–93] postnatal rat neural and glial cells,[94] as well as *Xenopus* oocytes[95] and *Drosophila* cells and tissues,[54,77,90,96] indicate that HSF1 resides in the nucleus both before and after heat stress. From these studies, it appears that nuclear translocation does not play a role in the activation of HSF1 during heat stress. Interestingly, in one study[79] done on nonstressed HeLa cells, HSF1 is found to reside predominantly in the cytoplasmic fraction using one biochemical fractionation procedure, while a different fractionation procedure shows HSF1 in both the nuclear and cytoplasmic fractions. This observation brings into question the reliability of localization results based on biochemical fractionation studies.[79] Mercier and co-workers[79] concluded that monomeric HSF1 in unstressed cells is a nuclear localized protein that has a relatively low affinity for DNA and that during fractionation, HSF1 diffuses out of the nuclei with the amount leaking out being dependent on the fractionation procedure and buffer used. Heat-stressed HSF1, on the other hand, has a high affinity for DNA and thus remains in the nuclear fraction.[79,90] The discrepancies in the immunocytochemical studies might be dependent on the source of anti-HSF1 antibodies as well as the particular cell type being examined (see Reference 79 for more discussion on these points).

The subnuclear distribution of HSF1 has also been the subject of several studies. It has been observed that upon heat shock in HeLa cells, HSF1 relocalizes in the nucleus to form large, irregularly shaped granules.[92,93,97] These HSF1 granules have been detected in immunofluorescence studies as well as in live HeLa cells using a GFP-HSF1 fusion protein. It is observed that the appearance and disappearance of the granules coincide with the activation and attenuation of HSF1.[92,97] In addition, the number of heat shock-induced HSF1 granules correlates with the ploidy of the cells, suggesting the presence of a specific chromosomal target for these granules.[93] Interestingly, however, despite the association of the granules with HSF1 activity, fluorescence *in situ* hybridization studies have indicated that these HSF1 foci are distinct from the sites of transcription of the HSP70 and HSP90 genes.[93] Treatment of HeLa cells with proteasome inhibitors leads to the formation of HSF1 granules that are noticeably different from those observed during heat shock, although the level of HSF1 DNA-binding activity and HSP70 transcription is comparable to that induced by heat shock.[98] In addition, only 15% of the cells treated with the proteasome inhibitors MG132 or *clasto*-lactacystin β-lactone display HSF1 granules, compared to 100% during heat shock and the granules are smaller and fewer per cell. It would appear that the transcriptional activity of HSF1 is not directly correlated with the presence, number, or morphology of the granules and, as a result, the roles of these HSF1 foci during the heat shock response remain elusive at present.

B.   Trimerization and DNA binding

Results from cross-linking assays and hydrodynamic studies indicate that under normal conditions, HSF1 exists as a latent, inactive monomer, which upon heat shock undergoes a monomer-to-trimer transition to acquire its high-affinity HSE-binding ability.[41,52–56] Trimerization is mediated by the HSF1 trimerization domain, which is composed of three arrays of hydrophobic heptad repeats that can form leucine zippers to allow HSF1 molecules to interact as a triple-stranded α-helical coiled coil[71] (see Figure 2.2). This trimerization is essential for the acquisition of high-affinity binding to HSE. Monomeric *Drosophila* HSF1 binds HSEs poorly with a binding constant (K_d) of 10^{-7} M.[99] Upon exposure to heat stress, the binding affinity of HSF1 increases by several orders of magnitude to a K_d of 10^{-13} M for *Drosophila* HSF[100] and 10^{-11} M for human HSF1.[101]

Studies in *Drosophila* and yeast have suggested that the overexpression of HSPs under normal conditions is disruptive to the growth and normal functioning of the organism.[102,103] As a result, it is essential that the activity of HSF1 be tightly regulated under normal conditions. Near the C-terminal end of HSF1, a region consisting of a fourth array of hydrophobic heptad repeats has been identified as a suppression domain for HSF1 trimerization.[72,73,76] Mutations of hydrophobic residues to charged amino acids in the fourth zipper motif of human HSF1 or the deletion of that region result in the constitutive trimerization and DNA-binding activity of HSF1 under normal conditions.[72,73,76] Based on these observations, a model has been proposed that describes the HSF1 monomer as a folded structure that is stabilized by the association of the N-terminal and C-terminal hydrophobic heptad repeats to form an intramolecular coiled coil, thus preventing the trimerization of HSF1 under nonstress conditions (Figure 2.2B).[72,73,76] In addition to the trimer suppression domain, a region near the N-terminus of HSF1 is also implicated in the negative regulation of HSF1 trimerization. Mutations in the short linker domain (amino acids 102–136 for human HSF1) between the DNA-binding domain and trimerization domain lead to constitutive trimerization and DNA binding of HSF1.[104] However, the precise role of the linker domain in the suppression of HSF1 oligomerization remains to be determined. Although it may have a direct regulatory function, it is conceivable that mutations in the linker domain may simply perturb the overall protein structure, thus destabilizing the HSF1 monomer.

How is the stress signal transduced to HSF1 to effect the conformational changes associated with a monomer-to-trimer transition during heat shock? In some studies, the overexpression of HSF1 in human cells leads to deregulated DNA-binding activity at nonstress conditions, leading to the suggestion that HSF1 oligomerization is controlled by a titratable negative regulatory factor.[41,76] In addition, expression of HSF1 in *Escherichia coli*, which presumably lacks the regulatory factors that may be present in human cells, results in a transcription factor that binds DNA in the absence of heat shock.[42] Furthermore, the activity of human HSF1 expressed in *Drosophila* cells is reprogrammed to adopt the induction temperature of the host cell, which is approximately 10°C lower than that observed in human cells.[105] Again, this can be interpreted as indirect evidence for the presence of cellular regulatory factors that mediate the activity of HSF1. What could this regulator be? Several lines of evidence favor the proposition that the heat shock proteins themselves act as regulators of HSF1 activity. Two possible candidates are HSP70 and HSP90, both of which have been shown to interact with HSF1.[81,85,86,106–109] More importantly, addition of exogenous HSP70 interferes with the *in vitro* activation of HSF1,[81] and overexpression of HSP70 *in vivo* leads to a decrease in DNA-binding activity of HSF1 under heat shock.[83,107] Likewise, immunodepletion of HSP90 from HeLa cell lysates or injection of anti-HSP90 antibodies into *Xenopus* oocytes results in a several-fold increase in HSF1-HSE binding activity, implicating the molecular chaperone in the negative regulation of HSF1 DNA-binding capacity.[86,109] Using the yeast two-hybrid protein interaction assay, Marchler and co-workers[110] have identified DroJ1 (*Drosophila* dnaJ/HSP40) as a protein that interacts with *Drosophila* HSF.

These results lead to the suggestion that the HSPs may interact with HSF1, perhaps to stabilize the intramolecular hydrophobic interactions within the HSF1 monomer; upon exposure to heat shock, the generation of nonnative proteins may compete away the HSPs from HSF1, thus allowing the transcription factor to become activated as the suppression is relieved.[83,86,109] In keeping with this model, many of the above-mentioned studies have proposed that "monomeric" HSF1 is stably associated with chaperones such as HSP70, HSP90, HSP40 (dnaJ), and other proteins found in chaperone complexes.[81,85,86,109] However, there is very little evidence to support the presence of such stable complexes *in vivo*, and at least some biochemical evidence indicates that the majority of inactive HSF exists as a true monomer.[47,56]

Contrasting the studies that support a role of HSPs in regulating HSF binding activity, a study from Rabindran and co-workers[106] indicates that the overexpression of HSP70/HSC70 does not affect the induction of HSF1 binding activity in Rat1 cells upon exposure to stress. Also contrasting the studies done in *Xenopus* and human cells,[86,109] depletion of HSP83 (HSP90) or DroJ1 individually in *Drosophila* cells using RNAi resulted in only a moderate effect on the activation of HSF1.[110] Interestingly, however, the simultaneous depletion of both HSC70 and DroJ1 resulted in a large proportion of the HSF in the cell being trimeric and transcriptionally active.[110] All of the observations discussed support a model in which at least three HSPs (70, 40, and 90) appear to have some role in maintaining HSF in its inactive form inside cells, although modulating the levels of any one of them may or may not affect HSF activity in a particular cell type (Figure 2.2B). This further suggests that there could be redundancy in the ability of the different HSPs to maintain HSF in the inactive form and that the requirement for a particular HSP for this function might be species and cell-type specific.

There are also a number of studies that have demonstrated that the trimerization and DNA-binding activity of HSF1 can be regulated in a temperature-responsive fashion in the absence of other protein factors.[111–115] When purified human[112] or mouse[111,113] HSF1 is exposed to heat *in vitro*, it is able to acquire trimerization and DNA-binding ability. Furthermore, purified *Drosophila* HSF displays reversible trimerization and DNA-binding activity when the temperature is elevated, returned to normal, and increased again.[114] With a view to these findings, an alternative model can be proposed in which the intrinsic structural characteristics of HSF1 may confer temperature sensitivity to the transcription factor based on heat-induced changes in intramolecular interactions that may destabilize the monomeric structure.[114,115] Although attractive, these *in vitro* data may not accurately reflect the *in vivo* conditions. For example, the *in vitro* heat-activated *Drosophila* HSF displays a lesser induction of trimerization and DNA binding in comparison to that observed after heat shock in cells.[114] Thus, it is conceivable that the intrinsic heat sensitivity of HSF1 may be additionally controlled by other cellular regulators, such as HSPs, that govern the activity of the transcription factor.

C. Transactivation

The activation of HSF1's transcriptional stimulatory properties involves at least a two-step mechanism in which HSF gains DNA-binding ability, followed by an apparently independently regulated acquisition of transactivation (transcriptional stimulatory) competence. Evidence that demonstrates the uncoupling of the two events can be obtained from studies characterizing the effects of various non-heat shock inducers of the heat shock response. For example, in the absence of heat, sodium salicylate is found to induce the DNA-binding activity of HSF1 in HeLa and *Drosophila* cells; however, this DNA-bound form of HSF1 is transcriptionally inert, as shown by its failure to induce HSP70 transcripts.[96,117] Similarly, hypertonic[117] and hypo-osmotic[118] stresses activate HSF1 DNA binding without transactivation. In another study,[76] overexpression of HSF1 in HeLa cells leads to constitutive DNA binding; however, transcriptional activation remains absent in this case. As a result, the acquisition of DNA-binding activity is not necessarily accompanied by the achievement of transactivational competence. However, there may be other explanations for the above observations. That is, certain inducers of the heat shock response may affect other processes in the cell besides the activation of HSF binding (e.g., transcription in general) and the amount of HSF within cells is normally tightly regulated and kept in its inactive form through an appropriate level of chaperones. That is, under "normal" conditions, HSF1 binding and transactivation properties may actually be coupled.

1. HSF1 repression domains

The uncoupling of DNA-binding activity and transactivation suggests that there may be independent mechanisms that control the two events, possibly involving different regulatory regions of HSF1 itself, or the interaction with or modification by, cellular regulatory factors. The transactivation domains of human and *Drosophila* HSF1 were identified in experiments that made chimeric fusion proteins which replaced the native HSF1 DNA-binding domain with the GAL4 DNA-binding domain of *Saccharomyces cerevisiae*. The resulting fusion proteins constitutively bind the GAL4 DNA-binding site and the transcriptional activity of various HSF deletion and point mutant constructs can be assayed directly.[74,75,77] From these studies, the transactivation domain, a region rich in acidic and hydrophobic residues, is mapped to the carboxy-terminal end of the protein, distal to amino acid residue 371 for human HSF1,[74] 395 for mouse HSF1,[75] and 603 for *Drosophila* HSF[77] (see Figure 2.2A). In some studies, the activation domain has been further subdivided into two units (AD1 and AD2) with distinct structures.[74,119] When the transactivation region alone is fused to the GAL4 DNA-binding domain, potent constitutive transcriptional activity is observed; in contrast, the chimeric factor that contains the rest of the regions N-terminal to the transactivation domain is essentially transcriptionally inert at control temperatures, but is induced upon heat shock to acquire transcriptional activity.[74,75] As a result, it appears that HSF1 transcriptional activity is negatively regulated at normal conditions and this regulation is dependent on a stress-responsive region distinct from the transactivation domain. Various studies have assigned this negative regulatory domain to the central region of human HSF1, spanning amino acids 221–310 in one study,[74] and in another study[76] to a region containing amino acids 228–276 in addition to part of the N–terminal leucine zippers (heptad repeats A/B). As well, a corresponding region consisting of amino acid residues 181–227 and part of the N-terminal leucine zippers is mapped in mouse HSF1.[75] The minimal negative regulation domains for mouse HSF1 (and presumably human in approximately the same places) occurs at amino acids 203–227 and 300–310 (Figure 2.2A).[32] The central HSF1 regions, when fused to the native HSF1 transactivation domain or the VP16 activation domain, are able to confer heat-inducible transcriptional activity that is independent of the mechanisms that regulate oligomerization and DNA binding.[74–76,119] It should be noted that the above-mentioned repression regions are distinct from the repression of trimerization domain (heptad repeat C in Figure 2.2A), which suppresses HSF activity by interacting with the oligomerization domain (heptad repeats A/B), preventing its trimerization.

2. Regulation by phosphorylation

Evidently, additional events apart from DNA-binding must occur for HSF1 to achieve full transcriptional potential. One mechanism that has been proposed involves phosphorylation/dephosphorylation events that may function as a switch to alter the activity of the transcription factor. The notion that inducible phosphorylation events may regulate HSF1 activity first came about from the observation that heat-activated HSF1 undergoes a decrease in electrophoretic mobility that can be reversed by phosphatase treatment.[36,41,120] In addition, ^{32}P-labeling experiments have shown that HSF1 from heat-shocked cells incorporates two- to threefold more ^{32}P over control levels.[78] Although hyperphosphorylation of HSF1 occurs during heat-induced HSF1 activation, it is unclear as to whether it plays a regulatory role in the activity of HSF1. Several lines of evidence suggest that, indeed, inducible phosphorylation is implicated in the acquisition of transcriptional competence by HSF1. For example, sodium salicylate has been found to induce trimerization and DNA-binding activity with the apparent absence of transcriptional activity;[96,116] interestingly, this transcriptionally inert HSF1 is not hyperphosphorylated, unlike that observed in the heat-activated transcription factor.[78] These correlations are contradicted by studies using the amino acid analog azetidine, which fails to induce HSF1 hyperphosphorylation[41] but is still capable of stimulating HSF1 DNA-binding activity and transcriptional activation. It has also been difficult to precisely identify phosphorylation sites on HSF1 that are phosphorylated only upon heat shock. A study on *Drosophila*

HSF by Fritsch and Wu[121] suggests that there are several phosphorylation sites on HSF that appear to incorporate more phosphates during heat shock. However, the authors of this study also observed that the steady-state phosphorylation of HSF remained unchanged after heat shock and that the phosphorylation of HSF1 plays no significant role in regulating its binding activity.[121] The study by Fritsch and Wu, however, did not directly address the role of HSF phopshorylation on transcriptional activation. A study with yeast (*Kluyveromyces lactis*) HSF, in which numerous serine and threonine residues were mutated to an amino acid that could not be phosphorylated, did not prevent the mutated HSF from stimulating heat shock gene transcription upon heat stress, suggesting that HSF hyperphosphorylation may not be required for HSF1 activity.[122] Thus, it is unclear at present whether the inducible phosphorylation observed during heat shock plays a regulatory role in HSF1 activity. One possibility is that the hyperphosphorylation of HSF1 observed during heat stress may simply reflect changes in the balance of cellular kinase and phosphatase activities, with some of these enzymes being affected more than others by increases in temperature.

In addition to inducible phosphorylation, constitutive phosphorylation has also been postulated to modulate the activity of HSF1. Several studies have indicated that constitutive phosphorylation at Ser303 and Ser307 in the negative regulatory domain suppresses the transcriptional activation of human HSF1[87–89] (see Figure 2.2A). These serine residues are found to be constitutively phosphorylated and mutating them to alanine to abolish their phosphorylation leads to derepression of HSF1 transcriptional activity in the absence of heat shock.[87–89] In the same studies, when the same serine residues are mutated to glutamine, which mimics phosphorylation, these mutated HSFs have normal heat inducibility. Although phosphorylation of Ser303 and Ser307 is implicated in the negative regulation of HSF1 activity under normal conditions, their dephosphorylation is not required for heat shock induction, suggesting that other regulatory mechanisms are involved in the transcriptional activation of HSF1.

A number of kinases have been implicated in the phosphorylation of Ser303 and Ser307. *In vitro* studies[123] have indicated that mitogen-activated protein kinase (MAPK) is the kinase responsible for phosphorylating Ser307, and that this event precedes the phosphorylation of Ser303 by glycogen synthase kinase 3 (GSK3).[89,124] The Raf/Erk MAP kinase pathway, and in particular Erk1, has been implicated in the negative regulation of HSF1 activity,[125] probably by increasing, either directly or indirectly, the amount of phosphates on Ser303 and Ser307. Inhibition of GSK3 activity in *Xenopus* results in increased HSF1 binding and heat shock gene transcription while overexpression of GSK3beta reduces these events, thus supporting the concept that GSK3 acts as a negative regulator of HSF1 activity.[126] All of these results imply that a number of cellular kinases, many of which are known to be involved in cell growth and proliferation, likely negatively regulate the basal activity of HSF1 under nonstress conditions.

V. Transcriptional activation of the heat shock genes by HSF1

How does HSF1 promote the transcriptional activation of heat shock genes? Several mechanisms have been proposed, based mainly on *in vitro* studies. Previous work examining the promoter region of the *Drosophila* HSP70 gene has shown that under noninduced conditions, RNA polymerase II is paused at a region between +20 and +30, poised to become activated upon heat shock.[127,128] This promoter-proximal transcriptional pause is also demonstrated at the human HSP70 promoter using permanganate footprinting analysis.[129] Instinctively, then, this may be a perfect place where the action of HSF1 can be elicited to relieve the transcriptional pause, thus activating the transcription of HSP70. Using *in vitro* chromatin assays, it has been shown that the disruption of nucleosome spacing at the *Drosophila* HSP70 promoter that occurs upon heat shock depends on a protein known as GAGA factor, as well as an ATP-dependent nucleosome remodeling factor.[130] This remodeling factor is known as NURF and is composed of at least four subunits, one of which has an ATPase domain and is called ISWI (imitation switch). ISWI is related but distinct from the SWI/SNF2 ATP-dependent chromatin remodeling complexes.[131,132] An *in vitro* study using human cells extracts has been able to reproduce the *in vivo* pausing phenomenon at a heterologous human HSP70 promoter containing GAL4 DNA-binding sites.[129] This study showed that the pausing is nucleosome dependent and can be relieved by the addition of a chimeric activator consisting of the GAL4 DNA-binding domain and the HSF1 transactivation domain. Together with a SWI/SNF remodeling complex, the HSF1 chimera is able to significantly increase the amount of full-length transcripts in the *in vitro* system.[129] Thus, it is conceivable that HSF1 might recruit chromatin-remodeling complexes to assist in nucleosome disruption, and in this way allow RNA polymerase II to enter into productive elongation. This model is consistent with the observation that HSF1 is associated with chromatin remodeling both near the promoter region and in the transcribed region of HSP70.[130]

In addition to the disruption of nucleosomes by ATP-dependent remodeling complexes, other modifications to chromatin may also be occurring. An immunocytochemical study that examined the distribution of phosphorylated and acetylated histone proteins on *Drosophila* polytene chromosomes found that while no appreciable change occurred in the distribution of acetylated histones H3 and H4 after heat shock, the staining pattern of phosphorylated histones was altered.[133] In particular, the total number of sites with phosphorylated H3 decreased during heat shock but at the same time phosphorylated H3 accumulated at the heat shock gene loci, suggesting a potential role for the phosphorylation of histones during activation of heat shock gene transcription.[133]

A different mechanism, although not necessarily exclusive to the above models, has been postulated for the role of HSF1 in the escape of RNA polymerase II from the transcriptional pause. Using affinity chromatography, *in vitro* binding assays, and co-immunoprecipitation analysis, HSF1 has been shown to interact with TATA-box binding protein (TBP) *in vitro* in *Drosophila* and both *in vitro* and *in vivo* in human.[134,135] Likewise, RNA polymerase II interacts with TBP via a C-terminal region (on RNA polymerase II) that possesses sequence similarity to acidic transactivators,[136,137] a class of activators to which HSF belongs. This interaction between RNA polymerase II and TBP, however, is disrupted upon addition of HSF, leading to the proposal of a "competition" model of HSF function.[134] Here, the pausing of RNA polymerase II at the HSP70 promoter is maintained by its interaction with TBP; upon activation of HSF during heat shock, HSF competes with RNA polymerase II for interaction with TBP, thus allowing the escape of Pol II into functional elongation. HSF may additionally promote progressive elongation by the recruitment of various elongation factors to the heat shock genes. To visualize the *in vivo* distribution of HSF and various elongation factors, immunocytochemical studies have been performed on the *Drosophila* polytene chromosomes. Thus far, the elongation factors that have been analyzed using this approach include P-TEFb, Spt5, and Spt6.[138–140] P-TEFb is implicated in the maturation of RNA Pol II into an elongationally competent complex[141,142] and Spt5 and Spt6 are shown to have a positive effect on transcriptional elongation.[143] Upon heat shock, Spt6 and P-TEFb are recruited to the heat shock gene loci and this recruitment is dependent on functional HSF activity.[138–140] Finally, there is some evidence to suggest that HSF may facilitate transcriptional reinitiation to promote multiple rounds of transcription that lead to the dramatic increase in expression of the HSPs during heat shock.[144]

VI. Attenuation of the heat shock response

In addition to the processes regulating the activation of HSF1, it is evident that mechanisms exist to down-regulate the heat shock response subsequent to its activation. This is suggested by the observation that the expression of heat shock genes attenuates despite the continuous presence of the heat stress.[145,146] The exposure of HeLa cells to continuous heat shock at 42°C results in the transcriptional activation of HSP70 up to 60 minutes, after which HSF dissociates from DNA and the transcription of the heat shock gene is attenuated.[80] Furthermore, the dissociation rate of HSF1 from HSE is tenfold faster *in vivo* than it is *in vitro*, suggesting the presence of an active process that facilitates dissociation.[80] Although the effect of HSP70 in the regulation of HSF1 activation is debatable, most studies agree that the chaperone plays a role in HSF1 deactivation. An autoregulatory role has been proposed for HSP70 which, following heat shock, accumulates to high levels in the cell and facilitates the deactivation of HSF1 in a negative feedback loop. This role of HSP70 is consistent with various observations. For example, the re-exposure of cells to a second heat shock following recovery from an initial heat shock induces a much lower level of HSP70 transcription and a more rapid attenuation response,[106,147] possibly as a result of the high level of HSPs that is already present. Moreover, when the expression of functional HSP70 is inhibited by the addition of actinomycin D, cycloheximide, or the arginine analog canavanine (which leads to the production of malfolded HSP70), the attenuation of HSF1 activity is either delayed or absent.[145,147] Conversely, the overexpression of HSP70 is associated with the accelerated loss of HSF DNA-binding activity and attenuation of HSP70 transcription following heat shock.[83,106,148] However, it is unclear whether the effect of HSP70 is exerted directly on HSF1, or if the accumulation of HSP70 is sensed (perhaps through the decrease in the level of nonnative proteins) and transduced to HSF1 via an indirect mechanism. Some evidence in support of a direct role of HSP70 arises from the demonstration that HSP70 interacts with active HSF1, in particular via the transactivation domain of the transcription factor.[81,85,106] Here, the overexpression of HSP70 is shown to inhibit HSF1 transcriptional activity, while the use of an HSP70 mutant that is deficient in chaperone function and is incapable of binding the HSF1 transactivation domain fails to affect transcription.[85] Studies by Ali and co-workers[109] suggest that HSP90 might also play a role in the attenuation of HSF1. In a two-hybrid assay with mouse HSF1, Satyal and co-workers[149] identified a novel interacting protein, which they named heat shock factor binding protein 1 (HSBP1), that may be involved in the deactivation of HSF1.[149] The 76-amino acid HSBP1 contains two arrays of hydrophobic repeats and can interact with the trimerization domain of activated HSF1 *in vitro*, but not the non-DNA binding form of the transcription factor. HSBP1 is also found in association with HSP70 during attenuation. Satyal and co-workers[149] have also demonstrated that HSBP1 can impair the DNA binding and transcriptional activity of HSF1; conceivably, HSBP1 can destabilize hydrophobic heptad interactions to facilitate the dissociation of the HSF1 trimer. Finally, some studies[122,150,151] contend that certain phosphorylation and dephosphorylation events on HSF1 may be involved in its deactivation by facilitating HSF–HSE dissociation, suppressing HSF transcriptional activity, and clearing HSF1 from the sites of transcription.

VII. Sensing and transduction of the stress signal

Studies from past years have accumulated a vast amount of knowledge in the processes and events that occur during HSF1 activation and attenuation. However, the precise stress sensing and transducing mechanism that signals to HSF1 for the onset of activation remains to be elucidated. What is the cellular sensor of stress? The heat shock response is induced by a myriad of environmental and chemical inducers, and the determination of their modes of action is central to the identification of the primary detector or effector of the stress stimulus. In the search for a unifying factor or common denominator for all of the inducers of the heat shock response, one hypothesis has been proposed that suggests that it is the appearance of abnormal proteins that triggers the stress cascade. Various lines of evidence are consistent with this proposal. First, most (if not all) of the stress inducers can be shown to either directly, or indirectly, result in the formation of nonnative proteins in the cell.[152] In addition, the injection of denatured proteins into *Xenopus* oocytes is sufficient to induce the activation of HSF1.[19,153] Also, the addition of glycerol, which stabilizes protein structure and thus buffers protein damage, leads to a decrease in HSF1 activation upon heat exposure *in vitro*.[154,155] Furthermore, different inducers have differing induction kinetics for HSF1 activity. Whereas heat rapidly activates HSF1, amino acid analogues confer a much more delayed activation, reflecting the time that is required for their incorporation into nascent polypeptides, and the resultant misfolding and accumulation of nonnative proteins in the cell.[145,146] Finally, it has been shown that proteasome inhibitors such as MG132 and lactacystin induce HSF1 activation in the absence of heat and that this effect is abolished by the protein synthesis inhibitor cycloheximide.[156–158] This can be interpreted as the consequence of the compromised ability of the proteasome to engage in the degradation and turnover of abnormal cellular proteins resulting from incorrect folding during or post translation. The addition of cycloheximide, then, may prevent the further build-up of abnormal proteins in the cell, thus obviating the activation of HSF1.

If abnormal proteins are the universal signal of cellular stress, then the sensor of the heat shock response must be able to detect their accumulation. A prime candidate, naturally, would be the molecular chaperones themselves. Here, the model suggests that HSP70, most likely in complexes with co-chaperones and other factors, participates in the maintenance of HSF1 in the inactive form under normal conditions. Upon heat shock or other stresses, denatured proteins accumulate and the demand for chaperones increases; this compromises their ability to maintain HSF1 in its inactive state, leading to its activation and the up-regulation of the heat shock genes. Indeed, the activation of HSF1 by the injection of denatured proteins into *Xenopus* oocytes[19,153] can be abolished by the co-injection of HSC70, which, conceivably, will allow the negative regulation of HSF1 to remain undisrupted.[159]

Alternatively or in addition to the above model, HSF1 may be capable of directly sensing the biochemical disturbances invoked by the stress agents. Studies in *Drosophila* have demonstrated that purified HSF can be directly and reversibly activated *in vitro* by heat and hydrogen peroxide.[114] Similarly, a decrease in pH in the physiological range induces the same activation of HSF[115] and this is relevant in that heat shock and various chemical inducers have been observed to decrease intracellular pH in living cells.[115,160,162] Thus, the intrinsic structure of HSF1 may allow it to directly sense changes in temperature, biochemical balance, and pH through stress-sensitive alterations in intramolecular interactions within the HSF1 molecule. For example, the trimerization domain and the C-terminal trimer suppression domain of HSF1 contain two histidine residues and more than 20 charged amino acids; a decrease in pH may disrupt the salt bridges and hydrogen bonds that stabilize the interactions that maintain the HSF1 monomer.[115] On the other hand, the notion that HSF1 may directly sense heat through temperature-induced changes in conformation probably cannot account for all the conditions under which HSF1 is activated. For example, some chemical inducers, such as cadmium sulfate, canavanine, and sodium arsenite, can induce the activation of HSF1 in the absence of heat and without a decrease in intracellular pH.[115] In addition, the temperature at which HSF1 activation is induced is not absolute. This is demonstrated by the observation that HeLa cells grown at a decreased culture temperature (35°C instead of 37°C) have a lower temperature requirement (40 to 41°C) for HSF1 activation.[80] More interestingly, the induction temperature of human HSF1 expressed in *Drosophila* cells is reset to that observed in the host.[105] Thus, it appears that the activation of HSF1 is dependent on the relative, rather than the absolute, change in environmental temperature. With a view to that, the elucidation of the precise mechanisms that mediate the sensing and transducing of the stress signal within the cell awaits further study.

VIII. Summary and perspectives

Through the integration of results obtained by the diverse experimental approaches employed thus far, a multi-step model of HSF1 activation begins to emerge. HSF1 is predominantly a nuclear protein and under nonstress conditions exists as a latent monomer whose DNA binding and transactivation activities are suppressed (see Figures 2.2B and 2.3). This suppression is achieved primarily by the interaction of the hydrophobic domain near the C-terminus of HSF1 with the trimerization domain found next to the DNA-binding domain in the N-terminal half of the protein. Cognate forms of the heat shock proteins/molecular chaperones likely play an important role in the formation and maintenance of this repressed state. During heat shock and other stresses, denatured proteins accumulate in the cell, necessitating the induction of the chaperone-mediated refolding or degradation of these proteins. Here, HSF1 becomes activated through conformational changes in its structure that arise via a direct self-sensing mechanism in response to changes in the physical environment and/or through the decrease in chaperone activity, causing HSF1 to go from its repressed (monomeric) to its active (trimeric) state. Trimerization of HSF1 fully exposes its DNA-binding domain, allowing for high-affinity binding to the HSEs in the heat shock gene promoters. HSF1 then achieves transcriptional competence through an independent process, possibly involving further conformational changes from self-detection of the physical environment and/or phosphorylation/dephosphorylation events. The transcriptionally competent HSF1 then activates the transcription of the heat shock genes by a mechanism that likely involves the recruitment of chromatin remodeling factors and the stimulation of RNA polymerase elongation and reinitiation. The products of the heat shock genes, after accumulating to a certain level within the cell, facilitate the inactivation of HSF1 in a negative feedback loop. Finally, HSF1 reverts back to its inactive monomeric form during the deactivation or attenuation phase of the heat shock response (Figure 2.3).

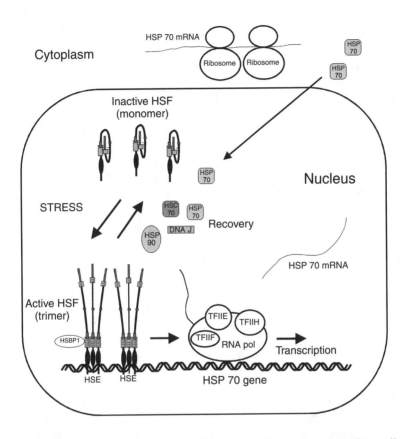

Figure 2.3 Model for the regulation of heat shock factor in mammalian cells. Inactive and active HSF reside in the nucleus. In response to heat and other forms of proteotoxic stress, inactive HSF monomers undergo a conformational change such that the interaction between the hydrophobic heptad regions found in the N and C terminal halves of the protein is disrupted promoting the formation of an HSF homotrimer. This conformational change may be a direct physical response of HSF to heat and/or it may be an indirect response to a decrease of HSP chaperone activity within the cell. HSF trimers bind heat shock elements (HSEs) found in the promoters of hs genes. HSF trimers likely undergo even further conformational changes to expose the transactivation domain found at the C-terminal of each HSF molecule. Active HSF stimulates the transcription of heat shock genes by releasing paused RNA polymerase II (RNA pol) and by recruitment of new polymerase complexes. Heat shock gene mRNAs are translated in the cytoplasm, thereby increasing the level of HSPs within the cell. Heat shock proteins (HSP) and heat shock cognate (HSC) proteins are translocated into the nucleus during heat shock and are thought to promote the refolding of active HSF back to its inactive monomeric form. In addition, another HSF interacting protein, HSBP1, is thought to bind to the hydrophobic heptad repeats of HSF1, thus contributing to its attenuation.

The heat shock response is an inducible defense mechanism that allows cells to cope with stressful environmental changes. Not only does it provide insight into how cells counteract and recover from stresses, but studies of the heat shock response also offer an excellent paradigm for investigating various aspects of transcriptional regulation in inducible gene expression systems. Moreover, the expression of heat shock proteins is found to increase during a number of pathophysiological states such as ischemia, cardiac hypertrophy, fever, inflammation, infection, aging, and cancer. Although it is unclear at present whether the HSPs are expressed as a cellular strategy to confer cytoprotective action during these disease states, or simply as a by-product of protein damage, studies of the regulation of the heat shock response can potentially offer enormous clinical significance.

Despite the tremendous advancement of knowledge over the years since the initial observation of heat shock-induced chromosomal puffs in *Drosophila*, much work remains to be done in the investigation of the regulation of the heat shock response. One challenge is to determine the relevance of the various proposed regulatory components, such as molecular chaperones, HSBP1, and phosphorylation events, in the multi-step activation and deactivation of HSF1. This feat can be approached by a combination of strategies. For example, the use of purified proteins to reproduce components of the heat shock response *in vitro* will allow the dissection of the heat shock pathway and the identification of direct effects conferred by one element to another; subsequently, further studies can be performed to confirm their *in vivo* significance. In addition, the determination of the mode of action of the various inducers of the heat shock response will permit a better understanding of the events and cellular conditions that lead to the activation of HSF1 under the seemingly diverse array of stress stimuli. Hopefully, future studies will allow the elucidation of the complex network of regulatory factors and molecular mechanisms that underlie the transcriptional regulation of the heat shock defense response in cells.

Acknowledgments

The authors would like to thank Jean Paul Paraiso for his assistance in the phylogenetic analysis of HSF proteins.

References

1. Storz, G. and Polla, B. S., Transcriptional regulators of oxidative stress-inducible genes in prokaryotes and eukaryotes, *Exs*, 77, 239, 1996.
2. Arrigo, A. P., Gene expression and the thiol redox state, *Free Radic. Biol. Med.*, 27, 936, 1999.
3. Allen, R. G. and Tresini, M., Oxidative stress and gene regulation, *Free Radic. Biol. Med.*, 28, 463, 2000.

4. Poulsen, H. E. et al., Antioxidants, DNA damage and gene expression, *Free Radic. Res.,* 33 (Suppl.), S33, 2000.

5. Zhu, Z. and Thiele, D. J., Toxic metal-responsive gene transcription, *Exs,* 77, 307, 1996.

6. Ghoshal, K. and Jacob, S. T., Regulation of metallothionein gene expression, *Prog. Nucleic Acid Res. Mol. Biol.,* 66, 357, 2000.

7. Miles, A. T. et al., Induction, regulation, degradation, and biological significance of mammalian metallothioneins, *Crit. Rev. Biochem. Mol. Biol.,* 35, 35, 2000.

8. Gething, M. J. and Sambrook, J., Protein folding in the cell, *Nature,* 355, 33, 1992.

9. Craig, E. A., Gambill, B. D., and Nelson, R. J., Heat shock proteins: molecular chaperones of protein biogenesis, *Microbiol. Rev.,* 57, 402, 1993.

10. Hendrick, J. P. and Hartl, F. U., Molecular chaperone functions of heat-shock proteins, *Annu. Rev. Biochem.,* 62, 349, 1993.

11. Georgopoulos, C. and Welch, W. J., Role of the major heat shock proteins as molecular chaperones, *Annu. Rev. Cell. Biol.,* 9, 601, 1993.

12. Parsell, D. A. and Lindquist, S., The function of heat-shock proteins in stress tolerance: degradation and reactivation of damaged proteins, *Annu. Rev. Genet.,* 27, 437, 1993.

13. Hartl, F. U., Molecular chaperones in cellular protein folding, *Nature,* 381, 571, 1996.

14. Pinto, M., Morange, M., and Bensaude, O., Denaturation of proteins during heat shock. *In vivo* recovery of solubility and activity of reporter enzymes, *J. Biol. Chem.,* 266, 13941, 1991.

15. Martin, J., Horwich, A. L., and Hartl, F. U., Prevention of protein denaturation under heat stress by the chaperonin HSP60, *Science,* 258, 995, 1992.

16. Parsell, D. A. et al., Protein disaggregation mediated by heat-shock protein HSP104, *Nature,* 372, 475, 1994.

17. Hightower, L. E., Cultured animal cells exposed to amino acid analogues or puromycin rapidly synthesize several polypeptides, *J. Cell. Physiol.,* 102, 407, 1980.

18. Levinson, W., Oppermann, H., and Jackson, J., Transition series metals and sulfhydryl reagents induce the synthesis of four proteins in eukaryotic cells, *Biochim. Biophys. Acta,* 606, 170, 1980.

19. Ananthan, J., Goldberg, A. L., and Voellmy, R., Abnormal proteins serve as eukaryotic stress signals and trigger the activation of heat shock genes, *Science,* 232, 522, 1986.

20. Westwood, J. T. and Steinhardt, R. A., Effects of heat and other inducers of the stress response on protein degradation in Chinese hamster and *Drosophila* cells, *J. Cell. Physiol.,* 139, 196, 1989.

21. Locke, M. et al., Activation of heat-shock transcription factor in rat heart after heat shock and exercise, *Am. J. Physiol.,* 268, C1387, 1995.

22. Polla, B. S. et al., Stress proteins in inflammation, *Ann. N.Y. Acad. Sci.,* 851, 75, 1998.

23. Amici, C. et al., Selective inhibition of virus protein synthesis by prostaglandin A1: a translational block associated with HSP70 synthesis, *J. Virol.,* 68, 6890, 1994.

24. Williams, R. S. and Benjamin, I. J., Protective responses in the ischemic myocardium, *J. Clin. Invest.,* 106, 813, 2000.

25. Ritossa, F., A new puffing pattern induced by temperature shock and DNP in *Drosophila, Experientia,* 18, 571, 1962.

26. Tissières, A., Mitchell, H. K., and Tracy, U. M., Protein synthesis in salivary glands of *Drosophila melanogaster*: relation to chromosome puffs, *J. Mol. Biol.*, 85, 389, 1974.
27. Lindquist, S., The heat-shock response, *Annu. Rev. Biochem.*, 55, 1151, 1986.
28. Nover, L., *Heat Shock Response*, CRC Press, Boca Raton, FL, 1991.
29. Lindquist, S. and Craig, E. A., The heat-shock proteins, *Annu. Rev. Genet.*, 22, 631, 1988.
30. Mager, W. H. and De Kruijff, A. J., Stress-induced transcriptional activation, *Microbiol. Rev.*, 59, 506, 1995.
31. Wu, C., Heat shock transcription factors: structure and regulation, *Annu. Rev. Cell. Dev. Biol.*, 11, 441, 1995.
32. Morimoto, R. I., Regulation of the heat shock transcriptional response: cross talk between a family of heat shock factors, molecular chaperones, and negative regulators, *Genes Dev.*, 12, 3788, 1998.
33. Nover, L. et al., The HSF world: classification and properties of plant heat stress transcription factors, *Cell Stress Chaperones*, 1, 215, 1996.
34. Morano, K. A. and Thiele, D. J., Heat shock factor function and regulation in response to cellular stress, growth, and differentiation signals, *Gene Expr.*, 7, 271, 1999.
35. Clos, J. et al., Molecular cloning and expression of a hexameric *Drosophila* heat shock factor subject to negative regulation, *Cell*, 63, 1085, 1990.
36. Sorger, P. K. and Pelham, H. R., Yeast heat shock factor is an essential DNA-binding protein that exhibits temperature-dependent phosphorylation, *Cell*, 54, 855, 1988.
37. Sarge, K. D. et al., Cloning and characterization of two mouse heat shock factors with distinct inducible and constitutive DNA-binding ability, *Genes Dev.*, 5, 1902, 1991.
38. Schuetz, T. J. et al., Isolation of a cDNA for HSF2: evidence for two heat shock factor genes in humans, *Proc. Natl. Acad. Sci. U.S.A.*, 88, 6911, 1991.
39. Nakai, A. et al., HSF4, a new member of the human heat shock factor family which lacks properties of a transcriptional activator, *Mol. Cell. Biol.*, 17, 469, 1997.
40. Nakai, A. and Morimoto, R. I., Characterization of a novel chicken heat shock transcription factor, heat shock factor 3, suggests a new regulatory pathway, *Mol. Cell. Biol.*, 13, 1983, 1993.
41. Sarge, K. D., Murphy, S. P., and Morimoto, R. I., Activation of heat shock gene transcription by heat shock factor 1 involves oligomerization, acquisition of DNA-binding activity, and nuclear localization and can occur in the absence of stress [published errata appear in *Mol. Cell. Biol.*, 13(5), 3122 and 13(6), 3838], *Mol. Cell. Biol.*, 13, 1392, 1993.
42. Rabindran, S. K. et al., Molecular cloning and expression of a human heat shock factor, HSF1, *Proc. Natl. Acad. Sci. U.S.A.*, 88, 6906, 1991.
43. Sarge, K. D. et al., Expression of heat shock factor 2 in mouse testis: potential role as a regulator of heat-shock protein gene expression during spermatogenesis, *Biol. Reprod.*, 50, 1334, 1994.
44. Mezger, V. et al., Heat shock factor 2-like activity in mouse blastocysts, *Dev. Biol.*, 166, 819, 1994.
45. Rallu, M. et al., Function and regulation of heat shock factor 2 during mouse embryogenesis, *Proc. Natl. Acad. Sci. U.S.A.*, 94, 2392, 1997.
46. Sistonen, L. et al., Activation of heat shock factor 2 during hemin-induced differentiation of human erythroleukemia cells, *Mol. Cell. Biol.*, 12, 4104, 1992.

47. Sistonen, L., Sarge, K. D., and Morimoto, R. I., Human heat shock factors 1 and 2 are differentially activated and can synergistically induce HSP70 gene transcription, *Mol. Cell. Biol.*, 14, 2087, 1994.

48. Tanabe, M. et al., The mammalian HSF4 gene generates both an activator and a repressor of heat shock genes by alternative splicing, *J. Biol. Chem.*, 274, 27845, 1999.

49. Goodson, M. L. and Sarge, K. D., Regulated expression of heat shock factor 1 isoforms with distinct leucine zipper arrays via tissue-dependent alternative splicing, *Biochem. Biophys. Res. Commun.*, 211, 943, 1995.

50. Goodson, M. L., Park-Sarge, O. K., and Sarge, K. D., Tissue-dependent expression of heat shock factor 2 isoforms with distinct transcriptional activities, *Mol. Cell. Biol.*, 15, 5288, 1995.

51. Fiorenza, M. T. et al., Complex expression of murine heat shock transcription factors, *Nucleic Acids Res.*, 23, 467, 1995.

52. Perisic, O., Xiao, H., and Lis, J. T., Stable binding of *Drosophila* heat shock factor to head-to-head and tail-to-tail repeats of a conserved 5 bp recognition unit, *Cell*, 59, 797, 1989.

53. Sorger, P. K. and Nelson, H. C., Trimerization of a yeast transcriptional activator via a coiled-coil motif, *Cell*, 59, 807, 1989.

54. Westwood, J. T., Clos, J., and Wu, C., Stress-induced oligomerization and chromosomal relocalization of heat-shock factor, *Nature*, 353, 822, 1991.

55. Baler, R., Dahl, G., and Voellmy, R., Activation of human heat shock genes is accompanied by oligomerization, modification, and rapid translocation of heat shock transcription factor HSF1, *Mol. Cell. Biol.*, 13, 2486, 1993.

56. Westwood, J. T. and Wu, C., Activation of *Drosophila* heat shock factor: conformational change associated with a monomer-to-trimer transition, *Mol. Cell. Biol.*, 13, 3481, 1993.

57. Pelham, H. R., A regulatory upstream promoter element in the *Drosophila* HSP 70 heat-shock gene, *Cell*, 30, 517, 1982.

58. Amin, J., Ananthan, J., and Voellmy, R., Key features of heat shock regulatory elements, *Mol. Cell. Biol.*, 8, 3761, 1988.

59. Xiao, H. and Lis, J. T., Germline transformation used to define key features of heat-shock response elements, *Science*, 239, 1139, 1988.

60. Fernandes, M., O'Brien, T., and Lis, J. T., Structure and regulation of heat shock gene promoters, in *The Biology of Heat Shock Proteins and Molecular Chaperones*, Morimoto, R. I., Tissières, A., and Georgopoulos, C., Eds., Cold Spring Harbor Laboratory Press, Cold Spring Harbor, NY, 1994.

61. Topol, J., Ruden, D. M., and Parker, C. S., Sequences required for *in vitro* transcriptional activation of a *Drosophila* HSP 70 gene, *Cell*, 42, 527, 1985.

62. Kroeger, P. E. and Morimoto, R. I., Selection of new HSF1 and HSF2 DNA-binding sites reveals difference in trimer cooperativity, *Mol. Cell. Biol.*, 14, 7592, 1994.

63. Wang, Y. and Morgan, W. D., Cooperative interaction of human HSF1 heat shock transcription factor with promoter DNA, *Nucleic Acids Res.*, 22, 3113, 1994.

64. Jedlicka, P., Mortin, M. A., and Wu, C., Multiple functions of *Drosophila* heat shock transcription factor *in vivo*, *EMBO J.*, 16, 2452, 1997.

65. McMillan, D. R. et al., Targeted disruption of heat shock transcription factor 1 abolishes thermotolerance and protection against heat-inducible apoptosis, *J. Biol. Chem.*, 273, 7523, 1998.

66. Xiao, X. et al., HSF1 is required for extra-embryonic development, postnatal growth and protection during inflammatory responses in mice, *EMBO J.*, 18, 5943, 1999.
67. Christians, E. et al., Maternal effect of HSF1 on reproductive success, *Nature*, 407, 693, 2000.
68. Harrison, C. J., Bohm, A. A., and Nelson, H. C., Crystal structure of the DNA binding domain of the heat shock transcription factor, *Science*, 263, 224, 1994.
69. Vuister, G. W. et al., Solution structure of the DNA-binding domain of *Drosophila* heat shock transcription factor [see comments], *Nat. Struct. Biol.*, 1, 605, 1994.
70. Vuister, G. W. et al., NMR evidence for similarities between the DNA-binding regions of *Drosophila melanogaster* heat shock factor and the helix-turn-helix and HNF-3/forkhead families of transcription factors, *Biochemistry*, 33, 10, 1994.
71. Peteranderl, R. and Nelson, H. C., Trimerization of the heat shock transcription factor by a triple-stranded alpha-helical coiled-coil, *Biochemistry*, 31, 12272, 1992.
72. Rabindran, S. K. et al., Regulation of heat shock factor trimer formation: role of a conserved leucine zipper, *Science*, 259, 230, 1993.
73. Zuo, J. et al., Activation of the DNA-binding ability of human heat shock transcription factor 1 may involve the transition from an intramolecular to an intermolecular triple-stranded coiled-coil structure, *Mol. Cell. Biol.*, 14, 7557, 1994.
74. Green, M. et al., A heat shock-responsive domain of human HSF1 that regulates transcription activation domain function, *Mol. Cell. Biol.*, 15, 3354, 1995.
75. Shi, Y., Kroeger, P. E., and Morimotor, R. I., The carboxyl-terminal transactivation domain of heat shock factor 1 is negatively regulated and stress responsive, *Mol. Cell. Biol.*, 15, 4309, 1995.
76. Zuo, J., Rungger, D., and Voellmy, R., Multiple layers of regulation of human heat shock transcription factor 1, *Mol. Cell. Biol.*, 15, 4319, 1995.
77. Wisniewski, J. et al., The C-terminal region of *Drosophila* heat shock factor (HSF) contains a constitutively functional transactivation domain, *Nucleic Acids Res.*, 24, 367, 1996.
78. Cotto, J. J., Kline, M., and Morimoto, R. I., Activation of heat shock factor 1 DNA binding precedes stress-induced serine phosphorylation. Evidence for a multistep pathway of regulation, *J. Biol. Chem.*, 271, 3355, 1996.
79. Mercier, P. A., Winegarden, N. A., and Westwood, J. T., Human heat shock factor 1 is predominantly a nuclear protein before and after heat stress, *J. Cell Sci.*, 112, 2765, 1999.
80. Abravaya, K., Phillips, B., and Morimoto, R. I., Attenuation of the heat shock response in HeLa cells is mediated by the release of bound heat shock transcription factor and is modulated by changes in growth and in heat shock temperatures, *Genes Dev.*, 5, 2117, 1991.
81. Abravaya, K. et al., The human heat shock protein HSP70 interacts with HSF, the transcription factor that regulates heat shock gene expression, *Genes Dev.*, 6, 1153, 1992.
82. Baler, R., Welch, W. J., and Voellmy, R., Heat shock gene regulation by nascent polypeptides and denatured proteins: HSP70 as a potential autoregulatory factor, *J. Cell Biol.*, 117, 1151, 1992.
83. Mosser, D. D., Duchaine, J., and Massie, B., The DNA-binding activity of the human heat shock transcription factor is regulated *in vivo* by HSP70, *Mol. Cell. Biol.*, 13, 5427, 1993.

84. Baler, R., Zou, J., and Voellmy, R., Evidence for a role of HSP70 in the regulation of the heat shock response in mammalian cells, *Cell Stress Chaperones*, 1, 33, 1996.

85. Shi, Y., Mosser, D. D., and Morimoto, R. I., Molecular chaperones as HSF1-specific transcriptional repressors, *Genes Dev.*, 12, 654, 1998.

86. Zou, J. et al., Repression of heat shock transcription factor HSF1 activation by HSP90 (HSP90 complex) that forms a stress-sensitive complex with HSF1, *Cell*, 94, 471, 1998.

87. Knauf, U. et al., Repression of human heat shock factor 1 activity at control temperature by phosphorylation, *Genes Dev.*, 10, 2782, 1996.

88. Kline, M. P. and Morimoto, R. I., Repression of the heat shock factor 1 transcriptional activation domain is modulated by constitutive phosphorylation, *Mol. Cell. Biol.*, 17, 2107, 1997.

89. Xia, W. et al., Transcriptional activation of heat shock factor HSF1 probed by phosphopeptide analysis of factor ^{32}P-labeled *in vivo*, *J. Biol. Chem.*, 273, 8749, 1998.

90. Wu, C. et al., Structure and regulation of the heat shock transcription factor, in *The Biology of Heat Shock Proteins and Molecular Chaperones*, Morimoto, R. I., Tissières, A., and Georgopoulos, C., Eds., Cold Spring Harbor Laboratory Press, Cold Spring Harbor, NY, 1994, 395.

91. Martinez-Balbas, M. A. et al., Displacement of sequence-specific transcription factors from mitotic chromatin, *Cell*, 83, 29, 1995.

92. Cotto, J., Fox, S., and Morimoto, R., HSF1 granules: a novel stress-induced nuclear compartment of human cells, *J. Cell Sci.*, 110, 2925, 1997.

93. Jolly, C. et al., HSF1 transcription factor concentrates in nuclear foci during heat shock: relationship with transcription sites, *J. Cell Sci.*, 110, 2935, 1997.

94. Brown, I. R. and Rush, S. J., Cellular localization of the heat shock transcription factors HSF1 and HSF2 in the rat brain during postnatal development and following hyperthermia, *Brain Res.*, 821, 333, 1999.

95. Mercier, P. A. et al., Xenopus heat shock factor 1 is a nuclear protein before heat stress, *J. Biol. Chem.*, 272, 14147, 1997.

96. Winegarden, N. A. et al., Sodium salicylate decreases intracellular ATP, induces both heat shock factor binding and chromosomal puffing, but does not induce HSP70 gene transcription in *Drosophila*, *J. Biol. Chem.*, 271, 26971, 1996.

97. Jolly, C., Usson, Y., and Morimoto, R. I., Rapid and reversible relocalization of heat shock factor 1 within seconds to nuclear stress granules, *Proc. Natl. Acad. Sci. U.S.A.*, 96, 6769, 1999.

98. Holmberg, C. I. et al., Formation of nuclear HSF1 granules varies depending on stress stimuli [In Process Citation], *Cell Stress Chaperones*, 5, 219, 2000.

99. Kim, S. J. et al., Interaction of the DNA-binding domain of *Drosophila* heat shock factor with its cognate DNA site: a thermodynamic analysis using analytical ultracentrifugation, *Protein Sci.*, 3, 1040, 1994.

100. Wu, C. et al., Purification and properties of *Drosophila* heat shock activator protein, *Science*, 238, 1247, 1987.

101. Taylor, I. C. et al., Facilitated binding of GAL4 and heat shock factor to nucleosomal templates: differential function of DNA-binding domains, *Genes Dev.*, 5, 1285, 1991.

102. Feder, J. H. et al., The consequences of expressing HSP70 in *Drosophila* cells at normal temperatures, *Genes Dev.*, 6, 1402, 1992.

103. Halladay, J. T. and Craig, E. A., A heat shock transcription factor with reduced activity suppresses a yeast HSP70 mutant, *Mol. Cell. Biol.*, 15, 4890, 1995.

104. Liu, P. C. and Thiele, D. J., Modulation of human heat shock factor trimerization by the linker domain, *J. Biol. Chem.*, 274, 17219, 1999.

105. Clos, J. et al., Induction temperature of human heat shock factor is reprogrammed in a *Drosophila* cell environment, *Nature*, 364, 252, 1993.

106. Rabindran, S. K. et al., Interaction between heat shock factor and HSP70 is insufficient to suppress induction of DNA-binding activity *in vivo*, *Mol. Cell. Biol.*, 14, 6552, 1994.

107. Baler, R., Zou, J., and Voellmy, R., Evidence for a role of HSP70 in the regulation of the heat shock response in mammalian cells, *Cell Stress Chaperones*, 1, 33, 1996.

108. Nair, S. C. et al., A pathway of multi-chaperone interactions common to diverse regulatory proteins: estrogen receptor, Fes tyrosine kinase, heat shock transcription factor HSF1, and the aryl hydrocarbon receptor, *Cell Stress Chaperones*, 1, 237, 1996.

109. Ali, A. et al., HSP90 interacts with and regulates the activity of heat shock factor 1 in *Xenopus* oocytes, *Mol. Cell. Biol.*, 18, 4949, 1998.

110. Marchler, G. and Wu, C., Modulation of *Drosophila* heat shock transcription factor activity by the molecular chaperone DROJ1, *EMBO J.*, 20, 499, 2001.

111. Goodson, M. L. and Sarge, K. D., Heat-inducible DNA binding of purified heat shock transcription factor 1, *J. Biol. Chem.*, 270, 2447, 1995.

112. Larson, J. S., Schuetz, T. J., and Kingston, R. E., *In vitro* activation of purified human heat shock factor by heat, *Biochemistry*, 34, 1902, 1995.

113. Farkas, T., Kutskova, Y. A., and Zimarino, V., Intramolecular repression of mouse heat shock factor 1, *Mol. Cell. Biol.*, 18, 906, 1998.

114. Zhong, M., Orosz, A., and Wu, C., Direct sensing of heat and oxidation by *Drosophila* heat shock transcription factor, *Mol. Cell*, 2, 101, 1998.

115. Zhong, M., Kim, S. J., and Wu, C., Sensitivity of *Drosophila* heat shock transcription factor to low pH, *J. Biol. Chem.*, 274, 3135, 1999.

116. Jurivich, D. A. et al., Effect of sodium salicylate on the human heat shock response, *Science*, 255, 1243, 1992.

117. Alfieri, R. et al., Activation of heat-shock transcription factor 1 by hypertonic shock in 3T3 cells, *Biochem. J.*, 319, 601, 1996.

118. Huang, L. E. et al., Rapid activation of the heat shock transcription factor, HSF1, by hypo-osmotic stress in mammalian cells, *Biochem. J.*, 307, 347, 1995.

119. Newton, E. M. et al., The regulatory domain of human heat shock factor 1 is sufficient to sense heat stress, *Mol. Cell. Biol.*, 16, 839, 1996.

120. Larson, J. S., Schuetz, T. J., and Kingston, R. E., Activation *in vitro* of sequence-specific DNA binding by a human regulatory factor, *Nature*, 335, 372, 1988.

121. Fritsch, M. and Wu, C., Phosphorylation of *Drosophila* heat shock transcription factor, *Cell Stress Chaperones*, 4, 102, 1999.

122. Hoj, A. and Jakobsen, B. K., A short element required for turning off heat shock transcription factor: evidence that phosphorylation enhances deactivation, *EMBO J.*, 13, 2617, 1994.

123. Chu, B. et al., Sequential phosphorylation by mitogen-activated protein kinase and glycogen synthase kinase 3 represses transcriptional activation by heat shock factor-1, *J. Biol. Chem.*, 271, 30847, 1996.

124. Chu, B. et al., Transcriptional activity of heat shock factor 1 at 37 degrees C is repressed through phosphorylation on two distinct serine residues by glycogen

synthase kinase 3 and protein kinases Calpha and Czeta, *J. Biol. Chem.*, 273, 18640, 1998.

125. Mivechi, N. F. and Giaccia, A. J., Mitogen-activated protein kinase acts as a negative regulator of the heat shock response in NIH3T3 cells, *Cancer Res.*, 55, 5512, 1995.

126. Xavier, I. J. et al., Glycogen synthase kinase 3beta negatively regulates both DNA-binding and transcriptional activities of heat shock factor 1, *J. Biol. Chem.*, 275, 29147, 2000.

127. Rougvie, A. E. and Lis, J. T., The RNA polymerase II molecule at the 5' end of the uninduced HSP70 gene of *D. melanogaster* is transcriptionally engaged, *Cell*, 54, 795, 1988.

128. Rasmussen, E. B. and Lis, J. T., *In vivo* transcriptional pausing and cap formation on three *Drosophila* heat shock genes, *Proc. Natl. Acad. Sci. U.S.A.*, 90, 7923, 1993.

129. Brown, S. A., Imbalzano, A. N., and Kingston, R. E., Activator-dependent regulation of transcriptional pausing on nucleosomal templates, *Genes Dev.*, 10, 1479, 1996.

130. Tsukiyama, T., Becker, P. B., and Wu, C., ATP-dependent nucleosome disruption at a heat-shock promoter mediated by binding of GAGA transcription factor, *Nature*, 367, 525, 1994.

131. Tsukiyama, T. and Wu, C., Purification and properties of an ATP-dependent nucleosome remodeling factor, *Cell*, 83, 1011, 1995.

132. Tsukiyama, T. et al., ISWI, a member of the SWI2/SNF2 ATPase family, encodes the 140 kDa subunit of the nucleosome remodeling factor, *Cell*, 83, 1021, 1995.

133. Nowak, S. J. and Corces, V. G., Phosphorylation of histone H3 correlates with transcriptionally active loci, *Genes Dev.*, 14, 3003, 2000.

134. Mason, P. B., Jr. and Lis, J. T., Cooperative and competitive protein interactions at the HSP70 promoter, *J. Biol. Chem.*, 272, 33227, 1997.

135. Yuan, C. X. and Gurley, W. B., Potential targets for HSF1 within the preinitiation complex [In Process Citation], *Cell Stress Chaperones*, 5, 229, 2000.

136. Xiao, H., Friesen, J. D., and Lis, J. T., A highly conserved domain of RNA polymerase II shares a functional element with acidic activation domains of upstream transcription factors, *Mol. Cell. Biol.*, 14, 7507, 1994.

137. Usheva, A. et al., Specific interaction between the nonphosphorylated form of RNA polymerase II and the TATA-binding protein, *Cell*, 69, 871, 1992.

138. Lis, J. T. et al., P-TEFb kinase recruitment and function at heat shock loci, *Genes Dev.*, 14, 792, 2000.

139. Andrulis, E. D. et al., High-resolution localization of *Drosophila* spt5 and spt6 at heat shock genes *in vivo*: roles in promoter proximal pausing and transcription elongation, *Genes Dev.*, 14, 2635, 2000.

140. Kaplan, C. D. et al., Spt5 and spt6 are associated with active transcription and have characteristics of general elongation factors in *D. melanogaster* [In Process Citation], *Genes Dev.*, 14, 2623, 2000.

141. Marshall, N. F. and Price, D. H., Purification of P-TEFb, a transcription factor required for the transition into productive elongation, *J. Biol. Chem.*, 270, 12335, 1995.

142. Peng, J. et al., Identification of multiple cyclin subunits of human P-TEFb, *Genes Dev.*, 12, 755, 1998.

143. Hartzog, G. A. et al., Evidence that Spt4, Spt5, and Spt6 control transcription elongation by RNA polymerase II in *Saccharomyces cerevisiae*, *Genes Dev.*, 12, 357, 1998.

144. Sandaltzopoulos, R. and Becker, P. B., Heat shock factor increases the reinitiation rate from potentiated chromatin templates, *Mol. Cell. Biol.*, 18, 361, 1998.

145. DiDomenico, B. J., Bugaisky, G. E., and Lindquist, S., The heat shock response is self-regulated at both the transcriptional and posttranscriptional levels, *Cell*, 31, 593, 1982.

146. Mosser, D. D., Theodorakis, N. G., and Morimoto, R. I., Coordinate changes in heat shock element-binding activity and HSP70 gene transcription rates in human cells, *Mol. Cell. Biol.*, 8, 4736, 1988.

147. Price, B. D. and Calderwood, S. K., Heat-induced transcription from RNA polymerases II and III and HSF binding activity are co-ordinately regulated by the products of the heat shock genes, *J. Cell. Physiol.*, 153, 392, 1992.

148. Kim, D., Ouyang, H., and Li, G. C., Heat shock protein HSP70 accelerates the recovery of heat-shocked mammalian cells through its modulation of heat shock transcription factor HSF1, *Proc. Natl. Acad. Sci. U.S.A.*, 92, 2126, 1995.

149. Satyal, S. H. et al., Negative regulation of the heat shock transcriptional response by HSBP1, *Genes Dev.*, 12, 1962, 1998.

150. Dai, R. et al., c-Jun NH_2-terminal kinase targeting and phosphorylation of heat shock factor-1 suppress its transcriptional activity, *J. Biol. Chem.*, 275, 18210, 2000.

151. Xia, W. and Voellmy, R., Hyperphosphorylation of heat shock transcription factor 1 is correlated with transcriptional competence and slow dissociation of active factor trimers, *J. Biol. Chem.*, 272, 4094, 1997.

152. Voellmy, R., Sensing stress and responding to stress, *Exs*, 77, 121, 1996.

153. Mifflin, L. C. and Cohen, R. E., Characterization of denatured protein inducers of the heat shock (stress) response in *Xenopus laevis* oocytes, *J. Biol. Chem.*, 269, 15710, 1994.

154. Edington, B. V., Whelan, S. A., and Hightower, L. E., Inhibition of heat shock (stress) protein induction by deuterium oxide and glycerol: additional support for the abnormal protein hypothesis of induction, *J. Cell. Physiol.*, 139, 219, 1989.

155. Mosser, D. D. et al., *In vitro* activation of heat shock transcription factor DNA-binding by calcium and biochemical conditions that affect protein conformation, *Proc. Natl. Acad. Sci. U.S.A.*, 87, 3748, 1990.

156. Bush, K. T., Goldberg, A. L., and Nigam, S. K., Proteasome inhibition leads to a heat-shock response, induction of endoplasmic reticulum chaperones, and thermotolerance, *J. Biol. Chem.*, 272, 9086, 1997.

157. Kawazoe, Y. et al., Proteasome inhibition leads to the activation of all members of the heat-shock-factor family, *Eur. J. Biochem.*, 255, 356, 1998.

158. Kim, D., Kim, S. H., and Li, G. C., Proteasome inhibitors MG132 and lactacystin hyperphosphorylate HSF1 and induce HSP70 and HSP27 expression, *Biochem. Biophys. Res. Commun.*, 254, 264, 1999.

159. Mifflin, L. C. and Cohen, R. E., hsc70 moderates the heat shock (stress) response in *Xenopus laevis* oocytes and binds to denatured protein inducers, *J. Biol. Chem.*, 269, 15718, 1994.

160. Drummond, I. A. et al., Large changes in intracellular pH and calcium observed during heat shock are not responsible for the induction of heat shock proteins in *Drosophila melanogaster*, *Mol. Cell. Biol.*, 6, 1767, 1986.

161. Weitzel, G., Pilatus, U., and Rensing, L., The cytoplasmic pH, ATP content and total protein synthesis rate during heat-shock protein inducing treatments in yeast, *Exp. Cell. Res.*, 170, 64, 1987.

162. Thompson, J. D. et al., The CLUSTAL_X windows interface: flexible strategies for multiple sequence alignment aided by quality analysis tools, *Nucleic Acids Res.*, 25, 4876, 1997.
163. Page, R. D., TreeView: an application to display phylogenetic trees on personal computers, *Comput. Appl. Biosci.*, 12, 357, 1996.

Addendum (*November 7, 2001*)

Note added in proof. Subsequent to the writing of this chapter, a review on heat shock factors has appeared (Pirkkala, L., Nykanen, P., and Sistonen, L., Roles of the heat shock transcription factors in regulation of the heat shock response and beyond, *FASEB J.*, 15(7), 1118–1131, 2001) as well as an article on the function of heat-induced HSF phosphorylation (Holmberg, C., Hietakangas, V., Mikhailov, A., Rantanen, J., Kallio, M., Meinander, A., Hellman, J., Morrice, N., MacKintosh, C., Morimoto, R. I., Eriksson, J. E., and Sistonen, L., Phosphorylation of serine 230 promotes inducible transcriptional activity of heat shock factor 1, *EMBO J.*, 20(14), 3800, 2001).

chapter three

Heat shock proteins and their induction with exercise

Earl G. Noble

Contents

I. Introduction

Striated muscle is a complex organ, which in man forms approximately 40% of total body mass. Constantly responding to an ever-changing environment, muscle can dramatically increase metabolism to sustain movement or it can be metabolized as a food source in times of decreased energy intake. Some muscles such as the heart are active on a continual basis, whereas others are

0-8493-0458-0/02/$0.00+$1.50
© 2002 by CRC Press LLC

only infrequently recruited. As a result, striated muscle is one of the most plastic and dynamic organs in the body.[1] Among the components that allow muscle to meet its many and varied tasks is a group of intracellular proteins known as heat shock proteins (HSPs) or stress proteins (SPs). In recent years, the wide range of activities performed by these proteins has begun to be elucidated and their actions fall into two main categories. They either maintain structural integrity or chaperone cellular constituents.[2] As a result, they play fundamental roles in protein synthesis,[3] degradation,[4] and stabilization[5] and are intimately involved in regulation of intracellular signaling.[6,7] HSPs are often associated with cellular protection and consequently are of great interest as potential therapeutic agents.[8,9] In this regard, exercise is unique in that it is a nontoxic, physiological inducer of HSPs.[10] With this in mind, the purpose of this chapter is to (1) review the major HSPs found in striated muscle; (2) comment on their response to exercise, and (3) identify some functional implications of this response. An extensive list of primary references is provided for the readers' use.

II. Major heat shock proteins in striated muscle

HSPs are a highly conserved superfamily of proteins that perform multiple, critical roles in both the developing and adult organisms.[9,11,12] Under conditions that result in the presence of partially denatured proteins (and possibly in response to other forms of signaling), heat shock elements (HSEs) located in the 5' region of HSP genes are bound by trimerized heat shock transcription factors (HSF1 usually, although others are involved) which upon phosphorylation or dephosphorylation result in varying degrees of transcriptional activation of heat shock genes.[13] This response is autoregulatory in that the resultant increase in HSP levels leads to a sequestration of HSF1 and termination of the stress response[14] (see also Chapter 2). There are several groups of HSPs, most often referred to on the basis of their molecular mass (i.e., the HSP70 family is comprised of proteins weighing in the 70-kDa range), that act independently and cooperatively to effect a concerted response to stress. The following, highlights a select list of the major HSPs found in striated muscle.

A.　Ubiquitin

Ubiquitin is a small-molecular-weight protein (8.5 kDa) that is part of the ubiquitin-proteasome pathway responsible for most of the non-lyosomal protein degradation, and hence most muscle protein degradation, in mammals. During this process, proteins that are no longer necessary or that cannot be rescued from unproductive folding are covalently bound by a branched polyubiquitin chain and are thereby targeted for degradation by the proteasome in the cytosol.[4] Proteins destined for ubiquitin targeting and destruction are identified by several means, including exposure of specific motifs such as may occur with protein unfolding or dissociation,[15,16] post-translational modifications such as phosphorylation[17] and dephosphorylation,[18] and interaction with other HSPs such as the 70-kDa heat shock protein cognate (constitutive isoform) HSC70.[19-21] In the later case, HSC70 may initially protect the protein, presenting it for degradation only after repeated attempts at refolding have been made.[19] Proteins may also be protected from degradation by the ubiquitin pathway by virtue of their quaternary structure. For example, intact myofibrils are not good substrates for ubiquitin-proteasome degradation, whereas damaged contractile proteins resulting from muscle injury are readily hydrolyzed.[22]

Recent evidence indicates an imp ortant role for ubiquitylation in cell signaling and regulation.[16,23-25] Mono as opposed to polyubiquitylation may target some membrane receptors, including several growth factor receptors, for degradation.[23,26] For example, when ubiquitin binds to growth hormone receptors, it can down-regulate these receptors via both lysosomal and proteasome-dependent pathways,[26,27] thereby directly affecting signal tranduction via the Jak/Stat pathway.[25] Moreover, the transcription factor MyoD, which is critical to muscle development, growth, and regeneration,[28] is degraded in the nucleus by the ubiquitin system unless it is bound (and presumably exerting transcriptional regulation) to its DNA-binding element.[16,24] Ubiquitin may exert an effect on protein synthesis by modulating accessibility to the DNA transcription complex[29] and by influencing the decay of cytokine and proto-oncogene mRNAs.[30] This ubiquitin targeting of growth factor receptors, muscle-specific transcription regulators, DNA accessibility, and mRNA viability is intriguing in that it suggests that ubiquitin-related systems might control protein turnover at both "input" and "output" stages.

B.   Small HSPs

The small HSPs are a superfamily of low-molecular-weight proteins (17 to 27 kDa), of which several members (HSP27 αB-crystallin, HSP20, MKBP/HSPB2, and HSPB3/cvHSP) are found in skeletal muscle.[31-39] As with other HSPs, these small HSPs exhibit dual roles that may not be mutually exclusive. Not only are they molecular chaperones[40-43] that have critical functions in development and maintenance of muscle,[5,32,44-47] but they also protect the muscle during periods of stress.[48-52]

HSP27 (HSP25) and αB-crystallin have been the most intensively investigated of the small HSPs but recent information suggests that others are also important. A 20-kDa HSP that is homologous to αB-crystallin binds to both αB-crystallin and HSP27.[33,34] MKBP/HSPB2, which interacts with myotonic dystrophy protein kinase (DMPK), maintains myofibril stability and is integral to the development of myotonic dystrophy.[39] HSPB3 is a recently characterized, small HSP whose precise function is unclear.[32,53] cvHSP, a 25-kDa protein, is selectively expressed in cardiovascular and insulin-sensitive tissues and may be involved in actin polymerization and associated with metabolism.[31] It has been suggested that the small HSPs can be divided into two functionally distinct groups.[32] HSP27, αB-crystallin and HSP20 (p20) are widely distributed in a variety of cells, including striated muscle, whereas MKBP/HSPB2 and HSPB3 are more localized to skeletal and, in the case of HSB3, cardiac muscle.[32,39,53,54] Moreover, in cell culture, Sugiyama et al.[32] observed that HSP27, αB-crystallin and HSP20 interacted with each other in various hetero- and homo-oligomeric configurations that were independent of MKBP/HSPB2 and HSPB3 oligomers. In addition to this selective inter-action, the small HSPs exhibit differential localization and developmental properties. HSP27 and αB-crystallin are up-regulated and bound to actin bundles in the early stages of myogenic differentiation while MKBP/HSPB2 is not.

As noted above, small HSPs normally exist in oligomeric aggregates of high molecular weight.[55,56] However, under conditions of development, or cellular stress, these proteins may assume smaller oligomers with potentially altered function as a result of phosphorylation.[34,56-58] For example, in devel-oping striated muscle, phosphorylated HSP27 is found at high levels,[5] whereas post-parturition, the degree of phosphorylation is reduced.[45,59] Given its cellular localization, at the I-band and M-line, it has been postulated that small phosphorylated HSP27 oligomers are associated with myofibril-logenesis and the insertion of actin into myofilaments.[5,40,60,61] The reduced phosphorylation observed with development would reflect a reduced rate of myofibrillogenesis while the continued high expression of HSP27 in oxi-dative fiber types compared to those that depend more on glycogenolysis probably reflects a higher rate of protein turnover in the former.[62]

HSP27 is also involved in microfilament stabilization during times of stress,[60,63] but here the effect of phosphorylation is more complex. Phosphory-lation of HSP27 may initially cause an increase in native molecular mass,[64,65] followed by a subsequent decrease in oligomeric size.[49,57,60,66,67] The large aggregates are critical for chaperone function[46,60,68] and may also increase the level of the antioxidant glutathione, thereby protecting cells against the dele-terious effects of oxygen free radicals.[68,69] The small oligomers are involved in microfilament stabilization and can enter the nucleus where they may either protect nuclear proteins or interact in cell signaling pathways.[70] In cells undergoing stress, the re-organization of HSP27 not only allows smaller oligomers to stabilize intracellular structures, but may enable them to retain chaperone activity and hold partially denatured proteins for subsequent refolding by HSC70.[49,50] HSP27 also binds specific initiation factors, thereby limiting mRNA transcription.[52] This might serve the effect of eliminating the production of proteins requiring chaperones, which are in short supply, having been recruited in an effort to maintain cellular viability.

αB-crystallin tends to co-localize with HSP27 at the I-band and M-line,[5,71] as well as at the Z-line in association with actin and desmin where it may help stabilize the Z-line.[72,73] Like HSP27, αB-crystallin is also involved in assembly and stabilization of the thin filament but possibly has an addi-tional involvement in intermediate filament regulation.[5] In this regard, indi-viduals exhibiting inherited, desmin-related dystrophies that included adult onset accumulation of desmin aggregates also displayed co-aggregation of an αB-crystallin gene defect. Further, muscle cell lines transfected with this mutant gene showed desmin and αB-crystallin aggregates that were similar to those observed in the patients.[74] αB-crystallin may also influence cytoskeletal remodeling by acting as a chaperone to present proteins to the proteasome for degradation.[75] HSP27, αB-crystallin increases post-parturition in both skeletal muscle, with the greatest elevations being observed in oxidative fibers.[45,76] The role of the altered phosphorylation state of αB-crystallin in these processes is unclear.[56]

There is very little information on the biological function of the other small HSPs. Changes in HSP20 during both development and in response to denervation suggest that HSP20 is influenced by neural input and is particularly important in frequently contracted oxidative muscles.[34,54] With stress, HSP20 either remains in the cytoplasm or migrates to the nucleus; but unlike HSP27 or αB-crystallin, it does not appear to be important in filament stabilization.[77]

C.   HSP32

HSP32 (or HO-1) is the inducible isoform of heme oxygenase. Along with the constitutively expressed isoform (HO-2), HSP32 is believed to act as an antioxidant by accelerating conversion of heme into biliverdin and ultimately bilirubin and carbon monoxide.[78] Heme, which is a pro-oxidant, can thus be converted into antioxidant metabolites.[79] In this way, HSP32 provides protection against atherosclerotic lesion and vascular injury[80,81] and is critical in guarding against myocardial ischemia–reperfusion injury.[82] Although HSP32 has a heat shock element in its promoter region,[83,84] it is most often up-regulated in response to oxidative stresses that may not operate through the traditional heat shock signaling pathways but may involve NFkB and AP-2 transcription factors.[85,86] Moreover, the increase in HSP32 expression in response to various oxidative stresses, including exercise,[82,87,88] does not appear to involve typical chaperone functions where specific proteins are stabilized or transported. Rather, HSP32 may provide protection through the aforementioned elimination of heme with coincident production of antioxidants[89] as well as the production of specific cellular messengers.[90,91] While still controversial,[91] carbon monoxide produced during the HO-1 production of biliverdin may be involved in this protective action, either directly or in conjunction with nitrous oxide.[90,92] Interestingly, hypoxia is also associated with increased HSP32 expression[93,94] but the importance of this under normal physiological conditions is unclear (see also Chapter 6).

D.   HSP47

HSP47, a glycoprotein located in the endoplasmic reticulum, is proposed to bind both collagen and procollagen molecules. It is involved in collagen processing and stabilizes procollagen in a monomeric state while chaperoning them to the Golgi apparatus.[95,96] This chaperone, which may function in a heteromeric complex with the glucose-regulated proteins (GRPs) 78 and 94, is responsive to various stressors, including heat shock.[97] HSP47 has also been found to bind with collagen associated with the basement membrane[98] and it is interesting to speculate as to whether this HSP might help maintain membrane integrity or be involved in specific signaling pathways during periods of stress.

E.   HSP70 family

Of the HSPs studied to date, the best characterized are those with an apparent molecular weight of 70 kDa. At least ten members of this family have been identified in mammals[99] and it is likely that each of the isoforms perform similar functions in different cellular compartments. As with HSP90,[100] HSP27,[101] and heme oxygenase,[90] cytoplasmic HSP70 is found in both constitutive (HSC70 or HSP73) and inducible (HSP70 or HSP72) isoforms that exhibit a plethora of functions. These include cell signaling,[6] mRNA stabilization and degradation,[30,102] protein degradation,[21] cytoskeletal interactions,[103,104] and even apoptosis or cell death.[105] However, more than any of the other HSP families, the HSP70s are involved in the chaperoning and folding of nascent polypeptides.[2,3,106] They interact with hydrophobic peptide segments of many newly synthesized proteins,[107,108] thereby preventing improper folding and aggregation of these polypeptides prior to their assumption of proper conformation. There may be as many as one HSP70 binding sequences per every 40 amino acids, and slight differences in motif may determine whether HSP70 acts to assist with protein folding or to chaperone proteins within the cell.[2] As a result, both HSC70 and HSP70 are integral to proteins synthesis, both through association with the ribosomal machinery[3,109–112] and direct involvement in regulation of translation initiation[113,114] (see also Chapter 4). Moreover, under conditions of protein denaturation or damage, members of the HSP70 family bind directly to these sequences to prevent protein aggregation[115,116] and, in conjunction with other HSPs, stimulate protein refolding. The N-terminal domain of the HSP70 family is a weak ATPase that appears to oscillate these proteins between strong and weak binding states. When ATP is present, protein recognition and binding occur; but the association is transient, and upon ATP hydrolysis (see accessory chaperones below), protein interaction is stabilized.[117] While they do not direct proteins to their final conformation, it is this sequential binding and release of the substrate that is important in stabilizing critical intermediates and preventing improper associations of elongating polypeptides.[2]

An important non-cytoplasmic member of the HSP70 family is the mitochondrial protein, HSP75 (GRP75).[118–120] (See also Chapter 8.) The mitochondrion is a unique organelle in that it carries DNA, which can code for many of its own constituents. However, several proteins must be imported from the cytoplasm and HSP75 is important in this regard.[12,121] Having been manufactured on ribosomes in the cytoplasm, proteins that have a specific mitochondrial targeting sequence[122] are chaperoned to the mitochondrial membrane by cytosolic HSP70s[123] and, in an energy-dependent process, HSP75 in conjunction with outer and inner mitochondrial translocase complexes transports the protein into the mitochondria.[119,124] Like HSC70, HSP75 may cycle between ADP- and ATP-bound configurations to accomplish mitochondrial protein import.[125,126] Once in the mitochondria, recently imported proteins are folded, assembled into enzymatic complexes, and subsequently released from an oligomeric ring structure of the mitochondrial chaperones HSP60 and HSP10.[2,121,127] This complex is important not only in chaperoning proteins that have been recently transported into the mitochondria, but also in refolding partially denatured proteins within mitochondria that have been subjected to stress.[128] Interestingly, a process involving HSC70 and a ring structure known as TRiC/CCT appears to perform a role similar to HSP60 in the cytoplasm.[129,130]

Recently, an HSP70-like family of proteins termed HSP110 has been described,[131] which is moderately expressed in heart and skeletal muscle.[132] Often found in large aggregates with HSP25 and HSC70,[133] HSP110 may provide thermotolerance and recovery from stress by holding unfolded proteins in preparation for refolding by HSP70,[134] much like HSP27.[49] The RNA binding properties of HSP110 suggest that, like HSP70, this protein could be involved in regulating mRNA stability or translation.[102]

Upon stress, both HSC70 and HSP70 redistribute within the cell[135,136] and, in doing so, protect the nucleus, nucleolus,[135–137] and cytoskeleton[138] and are important in helping maintain protein synthesis.[139] Stress-induced increases in HSP75 and associated proteins appear to protect mitochondria and enhance their capacity to maintain energy production during subsequent stressful events.[128,140,141] Cytosolic members of the HSP70 family may also protect the mitochondria[136,142] and, in conjunction with their role in the regulation of the stress-activated kinase JNK, may limit the degree of apoptosis.[105,143]

One of the more interesting questions with regard to the HSP70 family is why the cell has the nearly identical proteins HSC70 and HSP70 in the cytoplasm. In fact, in many tissues (including striated muscle), the so-called inducible isoform, HSP70, is found at relatively high constitutive levels.[37] Moreover, in response to a variety of stressors, including exercise, HSC70 is much less inducible than HSP70.[144] In attempting to address this issue, Brown and co-workers[145] observed that, following heat shock, HSP70 and HSC70 interacted with one another *in vivo* and co-localized in the nucleus and nucleolus and exhibited little difference in their biochemical characteristics.[145] More recently, it was demonstrated that although the localization of HSC70 and HSP70 were similar following stress (primarily in the nucleus and nucleoli and to some extent over the mitochondria), in resting cells, HSP72 had a primarily nuclear locus whereas HSC73 was distributed throughout the cell but concentrated over the mitochondria or in the nucleolus.[136] This suggests potentially different functions, or interaction with different accessory proteins in unstressed cells but common targets during stress. Further study will be required to address this issue.

F. HSP90 family

As with many of the other heat shock proteins, members of the HSP90 family exhibit differential compartmentation and activation but perform similar functions. Of these proteins, the two most important in vertebrates appear to be HSP90α and HSP90β, with HSP90β representing the constitutive iso-form and HSP90α the inducible isoform.[100] HSP90, which is very abundant in eukaryotic cells,[146] is generally present as homo-oligomers of various sizes[147,148] whose size increases in response to stress.[149] This oligomerization of HSP90 appears to be critical to its cellular function.[149] In an ATP-dependent cycle,[150,151] HSP90 will transiently interact with immature proteins either independently[152] or more usually as part of a multi-protein complex which frequently includes HSC70.[7,153] HSP90 is rapidly up-regulated in response to stress and like HSP27 acts to prevent the unwanted aggregation of partially unfolded proteins[154,155] by maintaining them in a folding compe-tent state for refolding by the HSC70 machinery.[156] Like HSP70, HSP90 may self-regulate its stress-induced synthesis through its interaction with the major heat shock transcription factor HSF1.[157–159]

HSP90 also exhibits more specific interactions within various signal transducing pathways which are critical to the adaptive capacity of the cell.[160,161] In this regard, the best described of the HSP interactions is that with steroid receptors in that HSP90 maintains the unstable receptors in a conformation that will allow them to interact with the target hormone and then transports the active complex to the nucleus where signal transduction can be executed.[7,153,162,163] In fact, the association of HSP90 with actin and other cytoskeletal elements[164,165] may not only represent a means of stabiliz-ing these structures, but also a means by which these signal transduction complexes are transported to and from the nucleus.[162] HSP90 appears to interact with other signaling pathways in an analogous manner, most notably with several transcription factors and protein kinases. For example, HSP90 is involved in the control of helix-loop-helix transcriptions factors, including the muscle-specific MyoD.[103,166,167] Not only does the transient interaction of HSP90 with MyoD convert this transcription factor to its active form,[167] but it also is critical in the assembly of functional MyoD oligmers with other transcription factors necessary for successful DNA binding and transcription.[166] Interestingly, during muscle development, it is the inducible HSP90α isoform that is co-localized with MyoD,[103,168,169] and in addition to increased expres-sion in response to stress, this isoform appears to play an essential role in muscle formation. The observation that HSP90 may help regulate both pro-tein synthesis[170] and degradation[162,171,172] supports its significance in muscle function.

Along with HSC70, HSP90 plays a prominent role in the mitogen-acti-vated signal cascade (MAP kinase).[6,7] The destabilization and subsequent degradation of tyrosine kinase receptors,[173,174] disrupted translocation of Raf to the cell membrane, and inhibition of MAP kinase activation[175–177] using the HSP90 inhibitory drug geldanamycin all point to the importance of HSP90 in a wide range of cellular signaling. Further, HSP90 appears to play an important role in nitrous oxide signaling[178] and is critical for proper vascular function.[179] Recently, a mitochondrial homologue of HSP90 has been characterized[180] but its importance to mitochondrial protection, signaling, and adaptation has yet to be determined.

G. Accessory proteins

In addition to the major HSPs noted above, there are several accessory chaperones that are critical for normal HSP function. Members of the HSP40 family stimulate HSP70 binding to substrates by increasing the ATPase activity of the HSP70s and HSC70-interacting protein (Hip) helps stabilize this complex, thereby enhancing chaperone function.[181–183] These chaperones, in conjunction with HSC70-HSP90 organizing protein (Hop), are not only important in regulating the function of individual HSPs, but are critical in the interaction of multiple HSPs in chaperone machines.[153,184] A co-chaperone identified as CHIP which attenuates HSC70 chaperone activity[171] likely assists HSC70 and HSP90 in targeting proteins for degradation via the ubiquitin-proteasome pathway, thereby linking several HSPs.[185,186] Moreover, Bag-1, which also negatively modulates HSP70 function,[183,187] exerts roles in both apoptotic[188] and intracellular signaling via Raf1/ERK pathways.[189] The above observations suggest that these accessory proteins (and numerous others that go beyond the scope of this review) not only modulate individual HSP activity but identify specific chaperone targets. In this manner, they can regulate whether HSP function is anabolic or catabolic, stimulatory or inhibitory. To date, the response to and action of these accessory proteins with physiological stressors such as exercise have yet to be examined.

III. Heat shock protein induction with exercise

Following exercise, the stress response can be invoked in an effort to maintain or regain cellular homeostasis. As has been detailed in several recent reviews[10,190–192] and addressed elsewhere in this text, exercise results in activation (and inactivation) of a number of cell signaling pathways, which vary in a muscle- and exercise-specific fashion.[193–195] Moderate changes in intracellular ATP concentrations with increases in ADP and AMP[196–198] decreased carbohydrate stores,[199] hypoxia,[200,201] ischemia,[202] and reduced intracellular pH[203,204] could all be responsible for exercise-induced expression of HSPs. Elevated exercise temperatures may result in partial protein denaturation,[205,206] mitochondrial uncoupling with subsequent reactive oxygen species (ROS) formation, and changes in membrane integrity[207–209] leading to HSP synthesis. Increases in intracellular calcium[210,211] and exercise-dependent release of hormones such as epinephrine and norepinephrine[212–214] may also result in HSP accumulation. Finally, direct exercise-induced muscle stretch[215,216] or damage resulting in cytokine release and initiation of the inflammatory response can alter HSP production.[217-219] The magnitude of the stress response and the role played by HSPs varies as a consequence of the exercise stimulus. Hence, the response to a single bout of exercise may represent a reaction designed to recover homeostasis, whereas the changes that occur with repetitive exercise training likely represent an adaption designed to maintain homeostasis.

A. Acute exercise

The first study to document the effect of exercise on the stress response was that conducted by Hammond and colleagues.[220] In this study, rats were swam to exhaustion and immediately post-exercise, hearts were extirpated and analyzed for changes in translation products of isolated mRNA. Although increased synthesis of HSP70 proteins were evident in hearts from animals that had undergone heat shock or aortic banding, no difference between control hearts and those from swimmers were observed. Subsequent studies employing treadmill running rats[88,209,214,221-231] and exercising humans[199,232-238] have demonstrated that acute exercise is a sufficient stressor to induce the synthesis of a number of HSP transcripts and/or proteins. Interestingly, there has been no further evaluation of the stress response to swimming with the exception of two abstracts (Pshedin et al., *Med. Sci. Sports Ex.*, 26, S134, 1994 and Kelly et al., *Med. Sci. Sports Ex.*, 27, S43, 1995), which resulted in conflicting observations regarding the potential of swimming as an inducer of HSP expression.

HSP70, the inducible isoform of the HSP70 family, has been the most extensively examined HSP with regard to its response to acute exercise. In animal models, HSP70 has generally been observed to rise in response to one to three bouts of exercise with the magnitude and timing of the increase being influenced by exercise intensity, core temperature achieved, and muscle examined. In response to moderate intensity exercise, Locke et al.[229] reported that activation and binding of HSF1 and increased expression of HSP70 mRNA were detected as early as 20 and 40 minutes, respectively, in cardiac muscle. Although these investigators did not examine skeletal muscle, the time course of activation was likely similar as increased HSP70 mRNA has been reported in the plantaris immediately upon cessation of exhaustive exercise.[209] Using a sensitive ELISA technique, increased protein levels have been detected by 30 minutes following exercise,[224] but most studies have examined the response at 24 to 48 hours post-activity where relative increases of 1 to 30-fold have been reported.[237]

New or enhanced synthesis of a number of other putative HSPs, including those of approximately 65, 90, and 100 kDa, has been reported in rats subjected treadmill running, but the actual proteins were not identified.[209,222] Induction of HSP32 in the rat tibialis anterior exhibits a time course similar to that noted above for HSP70 following electrical stimulation of a loaded muscle and is elevated sevenfold immediately post-exercise in the plantaris of rats that had run for 60 minutes.[88] With 15 minutes of acute electrical stimulation, HSP60 content was found to increase in mouse soleus (within 4 hours) but not in the EDL. Finally, as with other stressors,[239–241] HSC70 exhibits a minimal response to exercise.[223,231]

While the data from rodent exercise models indicate a rapid and dramatic HSP response to stress, in human models, the data are less clear. To date, HSF1 activation with exercise has not been examined in humans but during moderate intensity effort (60 to 70% VO_{2max}), HSP70 mRNA in skeletal muscle was found to have increased 40 minutes before the end of exhausting exercise and to follow a time course similar to that observed in rat cardiac muscle post-exercise.[199,229,234] However, up to 3 hours post-exercise, no increase in HSP70 was detected either in muscle[234] or lymphocytes.[242] Similarly, HSP60 and HSP70 were not found to increase in vastus lateralis in response to 45 minutes of singled-legged exercise at 70% VO_{2max} until 3 and 6 days, respectively, after the exercise bout.[237] In contrast, when previously trained subjects were subjected to a half-marathon, increases in HSPs27, 60, and 70 were detected in leucocytes immediately post-exercise, persisting in some subtypes up to 24 hours.[233] Moreover, although there was no change in transcription rate or accumulation of αB-crystallin in response to either exhaustive single-leg knee extension at 70% of the 2-minute maximal resistance or to a single 4-hour bout of cycling at 60% of VO_{2max}, HSP32 transcription and subsequent MRNA accumulation occurred very rapidly with the more intensive single-leg exercise.[236] Clearly, many factors, including HSP examined, training status,[243] exercise duration,[244,245] muscle or cell type examined,[223,231] and large inter-individual variability in induction of the stress response with activity[233,237,242,246] have an impact on HSP expression.

Changes in exercise intensity that lead to differences in muscle recruitment pattern and hence muscle loading also appear to have a significant effect on HSP expression.[247,248] In this regard, the HSP70 expression in various muscles known to undergo different recruitment patterns was recently examined 24 hours following a single bout of exercise (Milne and Noble, unpublished observations). Reminiscent of the manner in which muscle fiber cytochrome c content is elevated in relation to recruitment order,[249] HSP70 also exhibited a distinct recruitment effect, increasing in the more oxidative muscles at lower running speeds and increasing in the least oxidative muscles only when very intense exercise was performed (see Figure 3.1). Development of high muscle forces may also influence the time course and nature of HSP induction. Eccentric resistance exercise, which is known to lead to physical damage of the muscle,[250] results in a rapid stress response. If fact, in mouse muscle subjected to lengthening contractions, increases in HSP70 are evident within 6 hours post-exercise, peak at 1 to 3 days, and decline thereafter.[251] In humans, an increased level of polyubiquitin transcripts has been detected in lymphocytes of individuals that have undergone a weight-lifting circuit,[252] and an increase in ubiquitin[238] and HSP/HSC70 and HSP27[217] has been detected in human biceps 48 hours after a single bout of eccentric exercise.

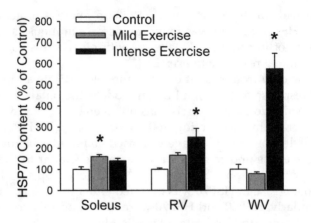

Figure 3.1   Relative HSP70 expression in rat muscle, 24 hours after running up a 2% grade at 15 (light exercise) or 30 (intense exercise) m/min for 60 minutes. Soleus and red and white portions of the vastus (RV and WV, respectively) were chosen because they represent muscles that are recruited in a sequential fashion with increasing running speed. HSP70 content is expressed as a percent of control ± SE; * indicates significantly greater than Con; $p < 0.05$; $n = 8$ per group.

B.   Exercise training

Elevated levels of several HSPs persist with repeated bouts of exercise despite adaptations that would tend to reduce the relative stress of exercise.[144,216,232,236,] [239,243–245,247,248,252–260] The myocardium exhibits up to a 12-fold increase in HSP70 following 8 to 10 weeks of treadmill run training at moderate exercise intensities in young rats (10 weeks of age).[247,256,261] However, αB-crystallin, HSP32, HSC70, HSP75, and HSP90 are unaltered following a similar training regimen at room temperature.[261] Using middle-aged animals (10 months old), Samelman[255] observed smaller (25 to 45%) but significant increases in both HSP70 (combined level of HSP70 and HSC70) and HSP60. In summary, these results suggest that chronic exercise training can maintain the increases in HSP70 in cardiac muscle observed following acute exercise.[214,224,225,228]

In skeletal muscle, exercise training is accompanied by an increase in most HSPs examined, including HSPs 60, 70, 75, and 78.[144,243,245,248,254,255,258,261,262] However, there are several reports in which the HSP70 content of skeletal muscle was unaltered following exercise training.[230,254,258] In two separate studies employing a training period of 8 weeks, training did not increases the content of HSPs70 or 60 and 75 in rat soleus.[254] In contrast, the mitochondrial chaperones (HSPs 60 and 75) were elevated in the less oxidative plantaris.[254] Muscles such as the soleus, which are rich in oxidative fiber types, do not tend to exhibit as great an HSP response to exercise training as more mixed or fast-twitch muscle,[144,243,254,255] despite the apparent capacity to do so.[263] Gonzalez et al.[243] observed an approximate doubling of HSPs 70, 75, and 78 in the soleus of rats subjected to an intense 3-month exercise training regimen yet in the less oxidative EDL, although HSPs75 and 78 demonstrated increases similar to those in the soleus, a much larger ninefold rise in HSP70 was observed.[243] Skeletal muscles with increased oxidative capacity generally exhibit a higher basal level of HSP70 (see below) and, as a consequence of this increased basal synthesis, the magnitude of the stress response may be attenuated.[264] Other factors such as age, gender (see Chapter 9), and strain (or race) may also play a critical role in the synthesis of HSPs in response to exercise.[221,227,265,266]

Most exercise studies examining the effect of exercise training on HSP expression have employed rodent models but a few studies have been conducted using human subjects.[232,233,245,248] Of these, none have compared trained individuals to a control group with the exception of one in which basal HSP levels in blood leucocytes of trained individuals were compared to untrained controls. Surprisingly, HSP27 and HSP70 levels were lower in the trained individuals, whereas no differences in the expression of HSC70 and HSP90 were observed.[233] In the same group of subjects, although the levels of HSP27 mRNA were down-regulated in the athletes, HSP70 mRNA expression was increased.[232] This suggests that an HSP-specific post-transcriptional regulation might be involved in the training response. In support of this, despite an increase in HSP70 mRNA having been observed in response to a single bout of exercise in humans,[234,237] Liu et al.[245] note that during an intense 4-week training period, the maximum level of HSP70 in highly trained rowers occurred about a week after the most intense training phase. These investigators and others[236] have speculated that in humans, there may be a time delay for HSP accumulation following exercise and that several bouts of activity causing transient increases in gene transcription may be required to elicit detectable change. Control of HSP expression at both transcription and translation has been reported[267] but whether it occurs in exercising humans has yet to be determined. In summary, it appears that exercise training results in a chronic increase in HSP70 in striated muscle. However, the magnitude of the response and the effect of training on other HSPs appear to be variable and regulated by a variety of factors.

One question that bears consideration is whether the accumulation of HSPs observed in skeletal muscle with exercise training blunts the ability of trained animals to respond to additional or novel stresses. In the most comprehensive study to date, Gonzalez and colleagues[243] examined the rates of synthesis of HSPs 70, 75, and 78 in rat soleus, 3 days after various durations of exercise training. At each time point, they measured resting HSP synthetic rates and their response to a single bout of exercise. They observed that with exercise training, the resting synthetic rate for all HSPs was reduced with a greater effect for HSPs 70 and 75 the longer the duration of the training. Nonetheless, the trained animals were responsive to a single bout of exercise 3 days after the last training session, exhibiting increased synthesis of all HSPs examined. In fact, when animals that had been trained for 3 months were given a single bout of exercise, the ratio of their post-exercise vs. resting synthetic rates for HSPs 27 and 70 were significantly elevated compared to sedentary controls.[243] Similar observations have been made with regard to HSP27 and 70 mRNA expression following heat shock when leucocytes from resting trained humans were compared to those from untrained controls.[232] These observations suggest that exercise training might allow individuals to mount a faster HSP response to external stressors while perhaps attenuating the magnitude of the increase accompanying training.

Whole body exercise is not required for a training effect as chronic electrical stimulation will elicit similar elevations in HSPs in striated muscle as those obtained with exercise.[246,268–270] Electrical stimulation is known to increase mitochondrial biogenesis,[119] hence the observed induction of members of the mitochondrial import machinery including HSPs 10, 60, 70, and 75 with electrical stimulation is likely associated with enhanced protein import into mitochondria.[269,271] Low-frequency chronic electrical stimulation is also associated with predictable changes in muscle fiber type with an increase in the more oxidative type I and type IIa fibers.[272] Coincident with this muscle transformation are increases in some of the HSPs found to be constitutively expressed in the more oxidative muscle fibers (see below), including ubiquitin, αB-crystallin, HSP27, and HSP70.[246,268,273] Ubiquitin levels are increased in rabbit skeletal muscle fibers that are undergoing necrosis (type IIx/d), as might be expected, but are also elevated in fibers that appear uninjured (IIa) and are simply transforming.[273]

C.   Fiber-specific expression

HSPs are not only rapidly induced in response to stress but are also constitutively expressed in many tissues, including striated muscle.[37] Within striated muscle there is a differential expression, however, in that muscle with a high proportion of oxidative fibers tends to exhibit elevated constitutive expression of ubiquitin,[274–277] HSP20,[54] αB-crystallin,[35,36,45,73,76,268,278] HSP27,[45,268] HSP70,[144,223,263,269,279] and HSPs 60 and 75.[269] At present, there is no information that HSP90 is distributed in a fiber-specific fashion; however, differences in the temporal expression of HSP90 in fast and slow muscle fibers has been noted during development[169] and, although not constitutively expressed, HSP32 induction is directly correlated with the percentage of oxidative fibers and myoglobin content.[280] In contrast, HSC70 appears to be constitutively expressed at similarly high levels in striated muscle of all types.[144,223] These data on whole muscle are supported by immunohistochemistry, which suggests that HSP70 and, αB-crystallin may be localized to the oxidative type I and IIa fibers.[76,246] Despite the general trend for elevated HSP expression in oxidative muscle, there are some notable exceptions. Cardiac muscle is the most oxidative of all striated muscle and although αB-crystallin, HSP27, and the mitochondrial chaperones HSPs 60 and 75 are found in high levels,[5,38,59,269] cardiac HSP70 content, at least in the rodent, is low compared to that observed in skeletal muscle such as the soleus.[223,229,279] Human myocardium has been reported to have high HSP70 levels,[281] but this may be explained by the fact that the tissue came from the hearts of patients undergoing surgery.[282]

The reason that cardiac muscle has low levels of HSP70 relative to oxidative fibers in skeletal muscle is unknown but it points to a specific role for HSP70 in the latter which need not be directly related to mitochondrial content and oxidative capacity.

Muscles rich in oxidative fiber types generally have increased roles in posture and ambulation, and elevated constitutive expression of HSPs could be a consequence of their greater use. It has been estimated that 2 to 5% of all oxygen taken up is released as reactive oxygen species (ROS)[283] and ROS are one by-product of frequent contractile activity.[231] These ROS may lead to membrane damage or protein oxidation and, hence, elevated levels of HSPs.[207,284] When muscle use is chronically reduced, such as occurs with hindlimb unweighting, denervation, or tenotomy, oxidative muscles such as the soleus demonstrate decreased levels of HSP70 and αB-crystallin.[45,285-287] Moreover, when soleus muscle is subjected to unloading as rats are run at increasing speeds,[249] HSP70 levels actually decline toward control values (Figure 3.1). This latter observation, which suggests that muscle loading rather than contractile activity per se is important, is in agreement with a previous report that demonstrated HSP32 was elevated to a greater degree in electrically stimulated muscle when the muscle was contracting against a load.[88] The response to muscle disuse is complex, however, as HSP27 is transiently increased in the soleus with denervation, whereas fast muscles such as the EDL demonstrate either no change or an increase in HSP27 and αB-crystallin, thereby suggesting that additional pathways beyond those which are neurally mediated may be important in the fiber-specific constitutive expression of HSPs.[45]

Differences in HSP expression in muscle with different fiber compositions could also be the consequence of their coordinated expression with other fiber-specific proteins. Fiber differences in protein composition are believed to be partially controlled by the differential expression of myogenic regulatory factors (MRFs).[288] At least one HSP (αB-crystallin) has regulatory elements in its promoter that not only interact with MRFs[289,290] but which in muscle undergoing continuous electrical stimulation are activated with a similar timecourse.[268] Under a variety of conditions, HSP70 expression is shifted in conjunction with slow or type I myosin heavy chain (type I MHC), which may be viewed as a marker for activation of the slow fiber type gene program. For example, when rat plantaris was subjected to compensatory overload in the presence of thyroid hormone, the normal overload induced increase of both type I MHC and HSP70 was blocked.[291] Moreover, the normal decline in HSP70 that occurs in the denervated soleus can be attenuated under a hypothyroid condition that also reduces the loss in type I MHC (O'Neill et al., unpublished observations). In individuals with spinal chord injuries who underwent an exercise training regimen, myosin mRNA levels for atrophied type II fibers were increased, while ubiquitin and 20 S proteasome mRNA levels, which are normally high in the more oxidative fibers,[277] declined.[292] These data support the contention that constitutive expression of HSPs may be partially under the control of factors specific to different muscle fiber types in addition to the direct effects of altered contractile activity. Experiments designed to elucidate these factors could potentially prove useful in offering muscle protection (see below) against various types of stress.

IV. Functional consequences of exercise-induced HSP expression

Studies employing transgenic animals provide direct evidence for the importance of HSPs, especially HSP70, to the protection of cardiac muscle against ischemia-reperfusion injury[128,293–299] despite studies that have questioned this relationship.[300,301] Moreover, exercise-induced elevations in cardiac HSP70 are correlated with better recovery from ischemic insult.[228,261] Since in cardiac muscle, chronic exercise training does not appear to be accompanied by a sustained increase in other HSPs,[247,261] this emphasizes the potential importance of exercise-induced HSP70 expression to cardioprotection. However, as noted above (see also Chapter 5), this response is complex because acute exercise in the cold can protect the heart during ischemia-reperfusion without an elevation in HSP70. In this instance, the investigators attributed the protection afforded by exercise to induction of antioxidant enzymes.[225] In a subsequent study involving exercise training (after which the activity of several antioxidant enzymes was unaltered), only hearts that exhibited a significant increase in HSP70 demonstrated protection from ischemia-reperfusion injury.[261] Interestingly, as previously mentioned, HSP70 induction may only occur with relatively intense exercise and it has been reported that the cardioprotective effects of exercise in humans are most influenced by the intensity of exercise.[302,303]

In contrast to the heart, a link between exercise-induced increases in HSPs and protection of skeletal muscle function has yet to be established. Using other pre-conditioning agents, such as heat shock, observations have been mixed. Enhanced survival of ischemia-reperfused skeletal muscle, with protection against mitochondrial injury and maintenance of creatine phosphate levels, have been reported following heat shock.[142,304–306] Heat stress has also been reported to attenuate skeletal muscle atrophy following hindlimb unweighting.[285] However, as with the heart, there are reports that question the role of HSPs in protecting skeletal muscle.[307,308]

Given the HSP response to exercise, it is reasonable to assume that these proteins may be involved in maintaining normal cellular homeostasis during and after exercise. For example, the damage caused to mitochondria during intense exercise[309] may be alleviated by HSP induction.[128] In one study, when rats were exercised at high environmental temperatures (36 to 37°C), those that had undergone heat shock (15 min at 41–42°C) 24 hours earlier were not only able to exercise longer (89 vs. 63 min), but also exhibited enhanced activity of selected mitochondrial enzymes involved in energy production.[140] HSP-mediated mitochondrial protection may take on a more critical role in muscle that has undergone severe, damaging exercise. Such exercise can lead to apoptosis or programmed cell death,[310] and leakage of cytochrome c from stressed mitochondria is believed to be one of the precipitating steps in this pathway.[311] The C-terminal N-terminal kinase (JNK) and p38 stress kinase pathways are key arbitrators of apoptosis[105] and they are activated in response to various stressors, including exercise,[193,312,313] especially if the contractions are more intense or exhibit an eccentric component.[314] Such contractions of course are those most likely to lead to muscle damage and cell death.[315] HSP70 modulates the JNK pathway by directly inhibiting JNK[105,316] or by interacting with other apoptosis-regulating proteins.[317,318] HSP27,[46,69] HSP60,[319,320] and αB-crystallin[321] may also be involved in regulating the apoptotic response to exercise. Whether the damaged muscle cell undergoes repair or apoptosis may ultimately depend on the availability of HSPs to stabilize and chaperone denatured proteins while simultaneously suppressing the activation of the stress kinases.

In addition to mitochondrial damage, damage to cell membranes such as the sarcolemma, the t-tubules, and the sarcoplasmic reticulum may be an outcome of exercise.[315] Reactive oxygen species are likely involved in this damage and may affect the ability of muscle to maintain force or recover from fatigue.[322,323] Given the potential chaperoning and stabilizing roles of HSPs, Thomas and Noble[324] subjected rat plantaris muscle to an *in situ* exercise protocol designed to induce a preferential reduction in force at low frequencies of stimulation (low-frequency fatigue). It was postulated that structural alterations in proteins, which are involved in calcium release and which give rise to low-frequency fatigue,[325,326] might be ameliorated by a priming stress (heat shock) designed to increase HSP levels. In fact, there is some evidence that HSP70 might modulate excitation contraction coupling[326] and HSP70s can interact with lipid membranes to form cation channels.[327] Although muscle HSP70 content (and presumably that of several other HSPs) was elevated in the heat-shocked animals, recovery from fatigue was not enhanced.[324] These results were confirmed in a subsequent study employing skeletal muscle from transgenic animals with up-regulated HSP70.[326]

HSPs are known to be involved in cytoskeletal restructuring during periods of stress and, when muscle is loaded, the cytoskeleton is rearranged.[328] Not only do HSPs help stabilize and chaperone elements of the cytoskeleton but, additionally, they are intimately involved in regulation of the altered protein turnover that occurs with exercise[329] (see above and also Chapter 4). In addition to their direct roles in transcription and translation, HSPs may play more indirect roles through their interaction with intracellular messengers. For example, both ubiquitin and HSP90 assist in the regulation of the MyoD, a transcription factor that is involved in growth and differentiation and whose expression is up-regulated in fast muscle fibers (see above). HSPs 70 and 90 bind to calmodulin and the serine-threonine phosphatase calcineurin, thereby stimulating calcineurin activity.[330,331] Calcineurin has recently been implicated in the regulation of both muscle fiber type profile[332,333] and muscle hypertrophy.[334,335] Interestingly, anabolic steroids increase the content of HSP70 in rat fast-twitch muscle, leading the authors to suggest that this could increase the exercise tolerance in these muscles.[259]

Other potential actions of HSPs induced in response to exercise remain speculative. Any process that requires maturation of protein complexes and intracellular chaperoning are clearly targets for HSP function. HSPs are also involved in the immune response and may be both up-regulated as a consequence of the immune response yet have a role in the regulation of cytokine expression[336] (see also Chapter 10). In this regard, the degree to which activation of the inflammatory response with exercise may confer protection beyond the exercised muscle (which had undergone damage) is unclear. In two studies that were conducted using single leg exercise, no stress response was noted in the contralateral muscle,[236,237] nor was HSP70 induced in the brain in response to exercise in the absence of cranial hyperthermia.[226] However, under conditions of induced pancreatitis, HSP70 was up-regulated in the lungs in response to circulating neutrophils.[337] Whether damaging exercise may similarly lead to systemic effects that might prove to be protective has not been fully evaluated but remains an intriguing possibility.

V. Summary and future directions

Exercise is one of the few physiologically relevant models to up-regulate HSP expression. Given their wide ranging influence on cell growth, adaptation, and protection, these proteins may play a vital role in the known benefits of exercise.[338] At the present time, however, there is little understanding of the molecular mechanisms underlying either the influence of exercise on HSP expression or the manner in which HSPs may regulate adaptation to activity. For example, members of the mitogen-activated protein kinase (MAPK) pathway(s), including the JNK, p38, and extracellular signal regulated protein kinases (ERKs), are implicated in the altered transcription patterns that occur post-exercise.[193,195,339] Not only do various HSPs interact with specific proteins in these pathways, thereby modulating their function,[6] but these kinases also influence HSP expression.[340–342] Similarly, the up-regulation of NOS[343] and cAMP,[344] which can occur in response to exercise, both regulates HSP expression[345,346] and is regulated by HSPs.[347] The role of these and other cell signaling pathways in mediating the stress response to exercise must be clarified. Moreover, only a select number of studies have simultaneously examined the response of several of the molecular chaperones to an exercise stress. To gauge the true impact of the stress response to exercise, such integration is necessary because HSPs often have individual cellular roles, yet under many circumstances condense to form heteromeric chaperone machines. The phosphorylation state of many of these chaperones is critical in this regard because phosphorylated HSPs may be acting individually at specific targets, performing disaggregation, renaturation, and stabilizing functions, whereas chaperone machines may be more engaged in higher-order functions such as signaling and cellular reconstruction.

In essence, the first response to exercise might be stabilization of a difficult situation, whereas the more long-term response might be mediation of molecular adaptation. Only by addressing such issues will we ultimately be able to judge the importance of the induction of HSPs with exercise.

Acknowledgments

This work was supported in part by grants from the National Sciences and Engineering and Research Council of Canada and the Heart and Stroke Foundation of Canada.

References

1. Booth, F. W. and Baldwin, K. M., Muscle plasticity: energy demand and supply processes, in *Exercise: Regulation and Integration of Multiple Systems*, Rowell, R. B. and Shepherd, J. T., Eds., Oxford University Press, New York, 1996, 1075.
2. Frydman, J., Folding of newly translated proteins *in vivo*: the role of molecular chaperones, *Annu. Rev. Biochem.*, 70, 603, 2001.
3. Beckmann, R. P., Mizzen, L. E., and Welch, W. J., Interaction of HSP 70 with newly synthesized proteins: implications for protein folding and assembly, *Science*, 248, 850, 1990.
4. Ciechanover, A., Orian, A., and Schwartz, A. L., Ubiquitin-mediated proteolysis: biological regulation via destruction, *Bioessays*, 22, 442, 2000.
5. Lutsch, G. et al., Abundance and localization of the small heat shock proteins HSP25 and αB-crystallin in rat and human heart, *Circulation*, 96, 3466, 1997.
6. Helmbrecht, K., Zeise, E., and Rensing, L., Chaperones in cell cycle regulation and mitogenic signal transduction: a review, *Cell. Prolif.*, 33, 341, 2000.
7. Pratt, W. B., The HSP90-based chaperone system: involvement in signal transduction from a variety of hormone and growth factor receptors, *Proc. Soc. Exptl. Biol. Med.*, 217, 420, 1998.
8. Jaattela, M., Heat shock proteins as cellular lifeguards, *Ann. Med.*, 31, 261, 1999.
9. Benjamin, I. J. and McMillan, D. R., Stress (heat shock) proteins: molecular chaperones in cardiovascular biology and disease, *Circ. Res.*, 83, 117, 1998.
10. Locke, M., The cellular stress response to exercise: role of stress proteins, *Exerc. Sport Sci. Rev.*, 25, 105, 1997.
11. Welch, W. J., Heat shock proteins functioning as molecular chaperones: their roles in normal and stressed cells, *Philos. Trans. R. Soc. Lond. B, Biol. Sci.*, 339, 327, 1993.
12. Fink, A. L., Chaperone-mediated protein folding, *Physiol. Rev.*, 79, 425, 1999.
13. Pirkkala, L., Nykanen, P., and Sistonen, L., Roles of the heat shock transcription factors in regulation of the heat shock response and beyond, *FASEB J.*, 15, 1118, 2001.
14. Shi, Y., Mosser, D. D., and Morimoto, R. I., Molecular chaperones as HSF1-specific transcriptional repressors, *Genes Dev.*, 12, 654, 1998.
15. Ciechanover, A., The ubiquitin-proteasome pathway: on protein death and cell life, *Embo J.*, 17, 7151, 1998.
16. Abu, H. O. et al., Degradation of myogenic transcription factor MyoD by the ubiquitin pathway *in vivo* and *in vitro*: regulation by specific DNA binding, *Mol. Cell. Biol.*, 18, 5670, 1998.

17. Brown, K. et al., Control of I kappa B-alpha proteolysis by site-specific, signal-induced phosphorylation, *Science*, 267, 1485, 1995.

18. Dimmeler, S. et al., Dephosphorylation targets Bcl-2 for ubiquitin-dependent degradation: a link between the apoptosome and the proteasome pathway, *J. Exp. Med.*, 189, 1815, 1999.

19. Bercovich, B. et al., Ubiquitin-dependent degradation of certain protein substrates *in vitro* requires the molecular chaperone HSC70, *J. Biol. Chem.*, 272, 9002, 1997.

20. Agarraberes, F. A., Terlecky, S. R., and Dice, J. F., An intralysosomal HSP70 is required for a selective pathway of lysosomal protein degradation, *J. Cell Biol.*, 137, 825, 1997.

21. Chiang, H. L. et al., A role for a 70-kilodalton heat shock protein in lysosomal degradation of intracellular proteins, *Science*, 246, 382, 1989.

22. Solomon, V. and Goldberg, A. L., Importance of the ATP-ubiquitin-proteasome pathway in the degradation of soluble and myofibrillar proteins in rabbit muscle extracts, *J. Biol. Chem.*, 271, 26690, 1996.

23. Hicke, L., Protein regulation by monoubiquitin, *Nat. Rev. Mol. Cell Biol.*, 2, 195, 2001.

24. Floyd, Z. E. et al., The nuclear ubiquitin-proteasome system degrades MyoD, *J. Biol. Chem.*, 276, 22468, 2001.

25. Strous, G. J. et al., Growth hormone-induced signal tranduction depends on an intact ubiquitin system, *J. Biol. Chem.*, 272, 40, 1997.

26. Shih, S. C., Sloper-Mould, K. E., and Hicke, L., Monoubiquitin carries a novel internalization signal that is appended to activated receptors, *EMBO J.*, 19, 187, 2000.

27. Alves dos Santos, C. M., van Kerkhof, P., and Strous, G. J., The signal transduction of the growth hormone receptor is regulated by the ubiquitin/proteasome system and continues after endocytosis, *J. Biol. Chem.*, 276, 10839, 2001.

28. Molkentin, J. D. and Olson, E. N., Combinatorial control of muscle development by basic helix-loop-helix and MADS-box transcription factors, *Proc. Natl. Acad. Sci. U.S.A.*, 93, 9366, 1996.

29. Mizzen, C. A. and Allis, C. D., Transcription. New insights into an old modification, *Science*, 289, 2290, 2000.

30. Laroia, G. et al., Control of mRNA decay by heat shock-ubiquitin-proteasome pathway, *Science*, 284, 499, 1999.

31. Krief, S. et al., Identification and characterization of cvHSP. A novel human small stress protein selectively expressed in cardiovascular and insulin-sensitive tissues, *J. Biol. Chem.*, 274, 36592, 1999.

32. Sugiyama, Y. et al., Muscle develops a specific form of small heat shock protein complex composed of MKBP/HSPB2 and HSPB3 during myogenic differentiation, *J. Biol. Chem.*, 275, 1095, 2000.

33. Kato, K. et al., Copurification of small heat shock protein with αB-crystallin from human skeletal muscle, *J. Biol. Chem.*, 267, 7718, 1992.

34. Kato, K. et al., Purification and characterization of a 20-kDa protein that is highly homologous to αB-crystallin, *J. Biol. Chem.*, 269, 15302, 1994.

35. Iwaki, T., Iwaki, A., and Goldman, J. E., αB-crystallin in oxidative muscle fibers and its accumulation in ragged-red fibers: a comparative immunohistochemical and histochemical study in human skeletal muscle, *Acta Neuropathol.*, 85, 475, 1993.

36. Iwaki, T., Kume-Iwaki, A., and Goldman, J. E., Cellular distribution of αB-crystallin in non-lenticular tissues, *J. Histochem. Cytochem.*, 38, 31, 1990.

37. Tanguay, R. M. and Khandjian, E. W., Tissue specific expression of heat shock proteins of the mouse in the absence of stress, *Dev. Genet.*, 14, 112, 1993.

38. Klemenz, R. et al., Expression of the murine small heat shock protein HSP25 and αB-crystallin in the absence of stress, *J. Cell Biol.*, 120, 639, 1993.

39. Suzuki, A. et al., MKBP, a novel member of the small heat shock protein family, binds and activates the myotonic dystrophy protein kinase, *J. Cell Biol.*, 140, 1113, 1998.

40. Benndorf, R. et al., Phosphorylation and supramolecular organization of murine small heat shock protein HSP25 abolish its actin polymerization-inhibiting activity, *J. Biol. Chem.*, 269, 20780, 1994.

41. Groenen, P. J. T. A. et al., Structure and modifications of the junior chaperone α-crystallin: from lens transparency to molecular pathology, *Eur. J. Biochem.*, 225, 1, 1994.

42. Horwitz, J., α-crystallin can function as a molecular chaperone, *Proc. Natl. Acad. Sci. U.S.A.*, 89, 10449, 1992.

43. Jakob, U. et al., Small heat shock proteins are molecular chaperones, *J. Biol. Chem.*, 268, 1517, 1993.

44. Benjamin, I. J. et al., Temporospatial expression of the small HSP/αB-crystallin in cardiac and skeletal muscle during mouse development, *Dev. Dyn.*, 208, 75, 1997.

45. Inaguma, Y. et al., Physiological and pathological changes in levels of the two small stress proteins, HSP27 and αB-crystallin, in rat hindlimb muscles, *J. Biochem.*, 114, 378, 1993.

46. Mehlen, P. et al., HSP27 as a switch between differentiation and apoptosis in murine embryonic stem cells, *J. Biol. Chem.*, 272, 31657, 1997.

47. Nover, L., Scharf, K. D., and Neumann, D., Cytoplasmic heat shock granules are formed from precursor particles and are associated with a specific set of mRNAs, *Mol. Cell. Biol.*, 9, 1298, 1989.

48. Loktionova, S. A. et al., Distinct effects of heat shock and ATP depletion on distribution and isoform patterns of human HSP27 in endothelial cells, *FEBS Lett.*, 392, 100, 1996.

49. Ehrnsperger, M. et al., Binding of non-native protein to HSP25 during heat shock creates a reservoir of folding intermediates for reactivation, *EMBO J.*, 16, 221, 1997.

50. Ehrnsperger, M. et al., The dynamics of HSP25 quaternary structure. Structure and function of different oligomeric species, *J. Biol. Chem.*, 274, 14867, 1999.

51. Perng, M. D. et al., Intermediate filament interactions can be altered by HSP27 and αB-crystallin, *J. Cell Sci.*, 112 (Pt. 13), 2099, 1999.

52. Cuesta, R., Laroia, G., and Schneider, R. J., Chaperone HSP27 inhibits translation during heat shock by binding eIF4G and facilitating dissociation of cap-initiation complexes, *Genes Dev.*, 14, 1460, 2000.

53. Boelens, W. C., van Boekel, M. A., and De Jong, W. W., HSPB3, the most deviating of the six known human small heat shock proteins, *Biochim. Biophys. Acta*, 1388, 513, 1998.

54. Inaguma, Y. et al., cDNA cloning of a 20-kDa protein (p20) highly homologous to small heat shock proteins: developmental and physiological changes in rat hindlimb muscles, *Gene*, 178, 145, 1996.

55. Rogalla, T. et al., Regulation of HSP27 oligomerization, chaperone function, and protective activity against oxidative stress/tumor necrosis factor alpha by phosphorylation, *J. Biol. Chem.*, 274, 18947, 1999.

56. Ito, H. et al., Phosphorylation-induced change of the oligomerization state of αB-crystallin, *J. Biol. Chem.,* 276, 5346, 2001.
57. Arrigo, A.-P. and Landry, J., Expression and function of the low-molecular weight heat shock proteins, in *Stress Proteins in Biology and Medicine,* Morimoto, R. I., Tissières, A., and Georgopoulos, C., Eds., Cold Spring Harbor Laboratory Press, Cold Spring Harbor, NY, 1994, 335.
58. Kantorow, M. and Piatigorsky, J., Phosphorylations of αA- and αB-crystallin, *Int. J. Biol. Macromol.,* 22, 307, 1998.
59. Gernold, M. et al., Development and tissue-specific distribution of mouse small heat shock proteins HSP25, *Dev. Genet.,* 14, 103, 1993.
60. Lavoie, J. N. et al., Modulation of cellular thermoresistance and actin filament stability accompanies phosphorylation-induced changes in the oligomeric structure of heat shock protein 27, *Mol. Cell. Biochem.,* 15, 505, 1995.
61. Miron, T. et al., A 25-kD inhibitor of actin polymerization is a low molecular mass heat shock protein, *J. Cell Biol.,* 114, 255, 1991.
62. Obinata, T. et al., Dynamic aspects of structural proteins in vertebrate skeletal muscle, *Muscle & Nerve,* 4, 456, 1981.
63. Li, S. et al., Fluid shear stress induces the phosphorylation of small heat shock proteins in vascular endothelial cells, *Am. J. Physiol.,* 271, C994, 1996.
64. Mehlen, P. et al., Tumor necrosis factor-alpha induces changes in the phosphorylation, cellular localization, and oligomerization of human HSP27, a stress protein that confers cellular resistance to this cytokine, *J. Cell Biochem.,* 58, 248, 1995.
65. Arata, S., Hamaguchi, S., and Nose, K., Inhibition of colony formation of NIH 3T3 cells by the expression of the small molecular weight heat shock protein HSP27: involvement of its phosphorylation and aggregation at the C-terminal region, *J. Cell Physiol.,* 170, 19, 1997.
66. Kantorow, M. and Piatigorsky, J., α-Crystallin/small heat shock protein has autokinase activity, *Proc. Natl. Acad. Sci. U.S.A.,* 91, 3112, 1994.
67. Kato, K. et al., Dissociation as a result of phosphorylation of an aggregated form of the small stress protein, HSP27, *J. Biol. Chem.,* 269, 11274, 1994.
68. Mehlen, P. et al., Large unphosphorylated aggregates as the active form of HSP27 which controls intracellular reactive oxygen species and glutathione levels and generates a protection against TNF-alpha in NIH-3T3-ras cells, *Biochem. Biophys. Res. Commun.,* 241, 187, 1997.
69. Baek, S. H. et al., Role of small heat shock protein HSP25 in radioresistance and glutathione-redox cycle, *J. Cell Physiol.,* 183, 100, 2000.
70. van de Klundert, F. A. et al., αB-crystallin and HSP25 in neonatal cardiac cells—differences in cellular localization under stress conditions, *Eur. J. Cell Biol.,* 75, 38, 1998.
71. Hoch, B. et al., HSP25 in isolated perfused rat hearts: localization and response to hyperthermia, *Mol. Cell. Biochem.,* 161/162, 231, 1996.
72. Bennardini, E., Wrzosek, A., and Chiesi, M., αB-crystallin in cardiac tissue: association with actin and desmin filaments, *Circ. Res.,* 71, 288, 1992.
73. Atomi, Y. et al., αB-crystallin in skeletal muscle: purification and localization, *J. Biochem.,* 110, 812, 1991.
74. Vicart, P. et al., A missense mutation in the αB-crystallin chaperone gene causes a desmin-related myopathy, *Nat. Genet.,* 20, 92, 1998.
75. Boelens, W. C., Croes, Y., and De Jong, W. W., Interaction between αB-crystallin and the human 20S proteasomal subunit C8/alpha7, *Biochim. Biophys. Acta,* 1544, 311, 2001.

76. Atomi, Y. et al., Fiber-type-specific αB-crystallin distribution and its shifts with T(3) and PTU treatments in rat hindlimb muscles, *J. Appl. Physiol.*, 88, 1355, 2000.

77. van de Klundert, F. A. and De Jong, W. W., The small heat shock proteins HSP20 and αB-crystallin in cultured cardiac myocytes: differences in cellular localization and solubilization after heat stress, *Eur. J. Cell Biol.*, 78, 567, 1999.

78. Stocker, R. et al., Bilirubin is an antioxidant of possible physiological importance, *Science Wash. D.C.*, 235, 1043, 1987.

79. Immenschuh, S. and Ramadori, G., Gene regulation of heme oxygenase-1 as a therapeutic target, *Biochem. Pharmacol.*, 60, 1121, 2000.

80. Duckers, H. J. et al., Heme oxygenase-1 protects against vascular constriction and proliferation, *Nat. Med.*, 7, 693, 2001.

81. Ishikawa, K. et al., Heme oxygenase-1 inhibits atherosclerotic lesion formation in ldl- receptor knockout mice, *Circ. Res.*, 88, 506, 2001.

82. Yoshida, T. et al., H(mox-1) constitutes an adaptive response to effect antioxidant cardioprotection: a study with transgenic mice heterozygous for targeted disruption of the heme oxygenase-1 gene, *Circulation*, 103, 1695, 2001.

83. Raju, V. S. and Maines, M. D., Coordinated expression and mechanism of induction of HSP32 (heme oxygenase-1) mRNA by hyperthermia in rat organs, *Biochim. Biophys. Acta*, 1217, 273, 1994.

84. Okinaga, S. and Shibahara, S., Identification of a nuclear protein that constitutively recognizes the sequence containing a heat-shock element. Its binding properties and possible function modulating heat-shock induction of the rat heme oxygenase gene, *Eur. J. Biochem.*, 212, 167, 1993.

85. Shibahara, S., Muller, R. M., and Taguchi, H., Transcriptional control of rat heme oxygenase by heat shock, *J. Biol. Chem.*, 262, 12889, 1987.

86. Tacchini, L. et al., Differential activation of heat-shock and oxidation-specific stress genes in chemically induced oxidative stress, *Biochem. J.*, 309 (Pt. 2), 453, 1995.

87. Applegate, L. A., Luscher, P., and Tyrrell, R. M., Induction of heme oxygenase: a general response to oxidant stress in cultured mammalian cells, *Cancer Res.*, 51, 974, 1991.

88. Essig, D. A., Borger, D. R., and Jackson, D. A., Induction of heme oxygenase-1 (HSP32) mRNA in skeletal muscle following contractions, *Am. J. Physiol.*, 272, C59, 1997.

89. Clark, J. E. et al., Heme oxygenase-1-derived bilirubin ameliorates postischemic myocardial dysfunction, *Am. J. Physiol. Heart Circ. Physiol.*, 278, H643, 2000.

90. Maines, M. D., The heme oxygenase system: a regulator of second messenger gases, *Annu. Rev. Pharmacol. Toxicol.*, 37, 517, 1997.

91. Cary, S. P. and Marletta, M. A., The case of CO signaling: why the jury is still out, *J. Clin. Invest.*, 107, 1071, 2001.

92. Vesely, M. J. et al., Heme oxygenase-1 induction in skeletal muscle cells: hemin and sodium nitroprusside are regulators *in vitro*, *Am. J. Physiol.*, 275, C1087, 1998.

93. Ryter, S. W. et al., Regulation of endothelial heme oxygenase activity during hypoxia is dependent on chelatable iron, *Am. J. Physiol. Heart Circ. Physiol.*, 279, H2889, 2000.

94. Heacock, C. S. and Sutherland, R. M., Enhanced synthesis of stress proteins caused by hypoxia and relation to altered cell growth and metabolism, *Br. J. Cancer*, 62, 217, 1990.

95. Smith, T. et al., HSP47 and cyclophilin B traverse the endoplasmic reticulum with procollagen into pre-Golgi intermediate vesicles. A role for HSP47 and cyclophilin B in the export of procollagen from the endoplasmic reticulum, *J. Biol. Chem.*, 270, 18323, 1995.

96. Satoh, M. et al., Intracellular interaction of collagen-specific stress protein HSP47 with newly synthesized procollagen, *J. Cell Biol.*, 133, 469, 1996.

97. Nagata, K., Saga, S., and Yamada, K. M., Characterization of a novel transformation-sensitive heat-shock protein (HSP47) that binds to collagen, *Biochem. Biophys. Res. Commun.*, 153, 428, 1988.

98. Ferreira, L. R. et al., HSP47 and other ER-resident molecular chaperones form heterocomplexes with each other and with collagen type IV chains, *Connect. Tissue Res.*, 33, 265, 1996.

99. Tavaria, M., Kola, I., and Anderson, R. L., The HSP70 genes of mice and men, in *Molecular Chaperones and Protein-Folding Catalysts,* Gething, M.-J., Ed., Oxford University Press, Oxford, 1997, 49.

100. Krone, P. H. and Sass, J. B., HSP90 alpha and HSP90 beta genes are present in the zebrafish and are differentially regulated in developing embryos, *Biochem. Biophys. Res. Commun.*, 204, 746, 1994.

101. Kawazoe, Y., Tanabe, M., and Nakai, A., Ubiquitous and cell-specific members of the avian small heat shock protein family, *FEBS Lett.*, 455, 271, 1999.

102. Henics, T. et al., Mammalian HSP70 and HSP110 proteins bind to RNA motifs involved in mRNA stability, *J. Biol. Chem.*, 274, 17318, 1999.

103. Sass, J. B., Weinberg, E. S., and Krone, P. H., Specific localization of zebrafish HSP90 alpha mRNA to myoD-expressing cells suggests a role for HSP90 alpha during normal muscle development, *Mech. Dev.*, 54, 195, 1996.

104. Sanchez, C. et al., Binding of heat-shock protein 70 (HSP70) to tubulin, *Arch. Biochem. Biophys.*, 310, 428, 1994.

105. Gabai, V. L. et al., Role of HSP70 in regulation of stress-kinase JNK: implications in apoptosis and aging, *FEBS Lett.*, 438, 1, 1998.

106. De Maio, A., The heat-shock response, *New Horizons*, 3, 198, 1995.

107. Gething, M. J. et al., Binding sites for HSP70 molecular chaperones in natural proteins, *Cold Spring Harb. Symp. Quant. Biol.*, 60, 417, 1995.

108. Flynn, G. C. et al., Peptide-binding specificity of the molecular chaperone BiP, *Nature*, 353, 726, 1991.

109. Eggers, D. K., Welch, W. J., and Hansen, W. J., Complexes between nascent polypeptides and their molecular chaperones in the cytosol of mammalian cells, *Mol. Biol. Cell*, 8, 1559, 1997.

110. Beck, S. C. and De Maio, A., Stabilization of protein synthesis in thermotolerant cells during heat shock, *J. Biol. Chem.*, 269, 21803, 1994.

111. Hansen, W. J., Cowan, N. J., and Welch, W. J., Prefoldin-nascent chain complexes in the folding of cytoskeletal proteins, *J. Cell Biol.*, 145, 265, 1999.

112. Nelson, R. J. et al., The translation machinery and 70 kd heat shock protein cooperate in protein synthesis, *Cell*, 71, 97, 1992.

113. Matts, R. L., Hurst, R., and Xu, Z., Denatured proteins inhibit translation in hemin-supplemented rabbit reticulocyte by inducing the activation of the heme-regulated eIF-2a kinase, *Biochemistry*, 32, 7323, 1993.

114. Vries, R. G. J. et al., Heat shock increases the association of binding protein-1 with initiation factor 4E, *J. Biol. Chem.*, 272, 32779, 1997.

115. Dobson, C. M. and Ellis, R. J., Protein folding and misfolding inside and outside the cell, *EMBO J.*, 17, 5251, 1998.

116. Gabai, V. L. and Kabakov, A. E., Rise in heat-shock protein level confers tolerance to energy deprivation, *FEBS Lett.,* 327, 247, 1993.

117. Zhu, X. et al., Structural analysis of substrate binding by the molecular chaperone DnaK, *Science,* 272, 1606, 1996.

118. Matouschek, A., Pfanner, N., and Voos, W., Protein unfolding by mitochondria. The HSP70 import motor, *EMBO Rep.,* 1, 404, 2000.

119. Hood, D. A., Invited Review: Contractile activity-induced mitochondrial biogenesis in skeletal muscle, *J. Appl. Physiol.,* 90, 1137, 2001.

120. Mizzen, L. A. et al., Identification, characterization, and purification of two mammalian stress proteins present in mitochondria, GRP75, a member of the HSP70 family and HSP58, a homolog of the bacterial groEL protein, *J. Biol. Chem.,* 264, 20664, 1989.

121. Bukau, B. and Horwich, A. L., The HSP70 and HSP60 chaperone machines, *Cell,* 92, 351, 1998.

122. Endo, T. et al., Binding of mitochondrial presequences to yeast cytosolic heat shock protein 70 depends on the amphiphilicity of the presequence, *J. Biol. Chem.,* 271, 4161, 1996.

123. Deshaies, R. J. et al., A subfamily of stress proteins facilitates translocation of secretory and mitochondrial precursor polypeptides, *Nature,* 332, 800, 1988.

124. Neupert, W., Protein import into mitochondria. [Review] [501 refs], *Annu. Rev. Biochem.,* 66, 863, 1997.

125. Horst, M. et al., Sequential action of two HSP70 complexes during protein import into mitochondria, *EMBO J.,* 16, 1842, 1997.

126. Azem, A. et al., The mitochondrial HSP70 chaperone system. Effect of adenine nucleotides, peptide substrate, and mGRPE on the oligomeric state of mHSP70, *J. Biol. Chem.,* 272, 20901, 1997.

127. Ostermann, J. et al., Protein folding in mitochondria requires complex formation with HSP60 and ATP hydrolysis, *Nature,* 341, 125, 1989.

128. Lau, S. et al., Simultaneous overexpression of two stress proteins in rat cardiomyocytes and myogenic cells confers protection against ischemia-induced injury, *Circulation,* 96, 2287, 1997.

129. Thulasiraman, V., Yang, C. F., and Frydman, J., *In vivo* newly translated polypeptides are sequestered in a protected folding environment, *EMBO J.,* 18, 85, 1999.

130. Frydman, J. et al., Folding of nascent polypeptide chains in a high molecular mass assembly with molecular chaperones, *Nature,* 370, 111, 1994.

131. Easton, D. P., Kaneko, Y., and Subjeck, J. R., The HSP110 and GRP1 70 stress proteins: newly recognized relatives of the HSP70s, *Cell Stress Chaperones,* 5, 276, 2000.

132. Lee-Yoon, D. et al., Identification of a major subfamily of large HSP70-like proteins through the cloning of the mammalian 110-kDa heat shock protein, *J. Biol. Chem.,* 270, 15725, 1995.

133. Wang, X. Y. et al., Characterization of native interaction of HSP110 with HSP25 and HSC70, *FEBS Lett.,* 465, 98, 2000.

134. Oh, H. J., Chen, X., and Subjeck, J. R., HSP110 protects heat-denatured proteins and confers cellular thermoresistance, *J. Biol. Chem.,* 272, 31636, 1997.

135. Welch, W. J. and Mizzen, L. A., Characterization of the thermotolerant cell. II. Effects on the intracellular distribution of heat-shock protein 70, intermediate filaments, and small nuclear ribonucleoprotein complexes, *J. Cell Biol.,* 106, 1117, 1988.

136. Ellis, S., Killender, M., and Anderson, R. L., Heat-induced alterations in the localization of HSP72 and HSP73 as measured by indirect immunohistochemistry and immunogold electron microscopy, *J. Histochem. Cytochem.,* 48, 321, 2000.

137. Welch, W. J. and Feramisco, J. R., Nuclear and nucleolar localization of the 72,000-dalton heat shock protein in heat-shocked mammalian cells, *J. Biol. Chem.,* 259, 4501, 1984.

138. Welch, W. J., Feramisco, J. R., and Blose, S. H., The mammalian stress response and the cytoskeleton: alterations in intermediate filaments, *Ann. N.Y. Acad. Sci.,* 455, 57, 1985.

139. Morcillo, G. et al., Specific intranucleolar distribution of HSP70 during heat shock in polytene cells, *Expt. Cell. Res.,* 236, 361, 1997.

140. Chen, H. W. et al., Previous hyperthermic treatment increases mitochondria oxidative enzyme activity and exercise capacity in rats, *Kaohsiung. J. Med. Sci.,* 15, 572, 1999.

141. Sammut, I. A. et al., Heat stress contributes to the enhancement of cardiac mitochondrial complex activity, *Am. J. Pathol.,* 158, 1821, 2001.

142. Bornman, L. et al., *In vivo* heat shock protects rat myocardial mitochondria, *Biochem. Biophys. Res. Commun.,* 246, 836, 1998.

143. Volloch, V. et al., Reduced thermo-tolerance in aged cells results from a loss of an HSP72- mediated control of JNK signaling pathway, *Cell Stress Chaperones,* 3, 265, 1998.

144. Kelly, D. A. et al., Effect of vitamin E deprivation and exercise training on induction of HSP70, *J. Appl. Physiol.,* 81, 2379, 1996.

145. Brown, C. R. et al., The constitutive and stress inducible forms of HSP70 exhibit functional similarities and interact with one another in an ATP-dependent fashion, *J. Cell Biol.,* 120, 1101, 1993.

146. Jakob, U., HSP90—News from the front, *Front Biosci.,* 1, d309, 1996.

147. Nemoto, T. and Sato, N., Oligomeric forms of the 90-kDa heat shock protein, *Biochem. J.,* 330 (Pt. 2), 989, 1998.

148. Garnier, C. et al., Phosphorylation and oligomerization states of native pig brain HSP90 studied by mass spectrometry, *Eur. J. Biochem.,* 268, 2402, 2001.

149. Nemoto, T. K., Ono, T., and Tanaka, K., Substrate-binding characteristics of proteins in the 90 kDa heat shock protein family, *Biochem. J.,* 354, 663, 2001.

150. Prodromou, C. et al., The ATPase cycle of HSP90 drives a molecular 'clamp' via transient dimerization of the N-terminal domains, *EMBO J.,* 19, 4383, 2000.

151. Panaretou, B. et al., ATP binding and hydrolysis are essential to the function of the HSP90 molecular chaperone *in vivo, EMBO J.,* 17, 4829, 1998.

152. Buchner, J. et al., GroE facilitates refolding of citrate synthase by suppressing aggregation, *Biochemistry,* 30, 1586, 1991.

153. Kimmins, S. and MacRae, T. H., Maturation of steroid receptors: an example of functional cooperation among molecular chaperones and their associated proteins, *Cell Stress Chaperones,* 5, 76, 2000.

154. Buchner, J., Supervising the fold: functional principles of molecular chaperones, *FASEB J.,* 10, 10, 1996.

155. Scheibel, T., Weikl, T., and Buchner, J., Two chaperone sites in HSP90 differing in substrate specificity and ATP dependence, *Proc. Natl. Acad. Sci. U.S.A.,* 95, 1495, 1998.

156. Freeman, B. C. and Morimoto, R. I., The human cytosolic molecular chaperones HSP90, HSP70 (HSC70) and HDJ-1 have distinct roles in recognition of a non-native protein and protein refolding, *EMBO J.,* 15, 2969, 1996.

157. Zou, J. et al., Repression of heat shock transcription factor HSF1 activation by HSP90 (HSP90 complex) that forms a stress-sensitive complex with HSF1, *Cell*, 94, 471, 1998.
158. Ali, A. et al., HSP90 interacts with and regulates the activity of heat shock factor 1 in *Xenopus* oocytes, *Mol. Cell. Biol.*, 18, 4949, 1998.
159. Bharadwaj, S., Ali, A., and Ovsenek, N., Multiple components of the HSP90 chaperone complex function in regulation of heat shock factor 1 *in vivo*, *Mol. Cell. Biol.*, 19, 8033, 1999.
160. Rutherford, S. L. and Lindquist, S., HSP90 as a capacitor for morphological evolution, *Nature*, 396, 336, 1998.
161. Freeman, B. C. and Yamamoto, K. R., Continuous recycling: a mechanism for modulatory signal transduction, *Trends Biochem. Sci.*, 26, 285, 2001.
162. Pratt, W. B. and Toft, D. O., Steroid receptor interactions with heat shock protein and immunophilin chaperones, *Endocrine Rev.*, 18, 306, 1997.
163. Czar, M. J. et al., Geldanamycin, a heat shock protein 90-binding benzoquinone ansamycin, inhibits steroid-dependent translocation of the glucocorticoid receptor from the cytoplasm to the nucleus, *Biochemistry*, 36, 7776, 1997.
164. Koyasu, S. et al., Two mammalian heat shock proteins, HSP90 and HSP100, are actin-binding proteins, *Proc. Natl. Acad. Sci. U.S.A.*, 83, 8054, 1986.
165. Czar, M. J., Welsh, M. J., and Pratt, W. B., Immunofluorescence localization of the 90-kDa heat-shock protein to cytoskeleton, *Eur. J. Cell Biol.*, 70, 322, 1996.
166. Shue, G. and Kohtz, D. S., Structural and functional aspects of basic helix-loop-helix protein folding by heat-shock protein 90, *J. Biol. Chem.*, 269, 2707, 1994.
167. Shaknovich, R., Shue, G., and Kohtz, D. S., Conformational activation of a basic helix-loop-helix protein (MyoD1) by the C-terminal region of murine HSP90 (HSP84), *Mol. Cell. Biol.*, 12, 5059, 1992.
168. Sass, J. B. and Krone, P. H., HSP90α gene expression may be a conserved feature of vertebrate somitogenesis, *Expt. Cell. Res.*, 233, 391, 1997.
169. Sass, J. B., Martin, C. C., and Krone, P. H., Restricted expression of the zebrafish HSP90alpha gene in slow and fast muscle fiber lineages, *Int. J. Dev. Biol.*, 43, 835, 1999.
170. Matts, R. L. et al., Interactions of the heme-regulated eIF-2 alpha kinase with heat shock proteins in rabbit reticulocyte lysates, *J. Biol. Chem.*, 267, 18160, 1992.
171. Ballinger, C. A. et al., Identification of CHIP, a novel tetratricopeptide repeat-containing protein that interacts with heat shock proteins and negatively regulates chaperone functions, *Mol. Cell. Biol.*, 19, 4535, 1999.
172. Connell, P. et al., The co-chaperone CHIP regulates protein triage decisions mediated by heat-shock proteins, *Nat. Cell Biol.*, 3, 93, 2001.
173. Miller, P. et al., Depletion of the erbB-2 gene product p185 by benzoquinoid ansamycins, *Cancer Res.*, 54, 2724, 1994.
174. Sato, S., Fujita, N., and Tsuruo, T., Modulation of Akt kinase activity by binding to HSP90, *Proc. Natl. Acad. Sci. U.S.A.*, 97, 10832, 2000.
175. Wartmann, M. and Davis, R. J., The native structure of the activated Raf protein kinase is a membrane-bound multi-subunit complex, *J. Biol. Chem.*, 269, 6695, 1994.
176. Schulte, T. W., An, W. G., and Neckers, L. M., Geldanamycin-induced destabilization of Raf-1 involves the proteasome, *Biochem. Biophys. Res. Commun.*, 239, 655, 1997.
177. Schulte, T. W. et al., Destabilization of Raf-1 by geldanamycin leads to disruption of the Raf- 1-MEK-mitogen-activated protein kinase signalling pathway, *Mol. Cell. Biol.*, 16, 5839, 1996.

178. Song, Y., Zweier, J. L., and Xia, Y., Heat-shock protein 90 augments neuronal nitric oxide synthase activity by enhancing Ca^{2+}/calmodulin binding, *Biochem. J.,* 355, 357, 2001.
179. Garcia-Cardena, G. et al., Dynamic activation of endothelial nitric oxide synthase by HSP90, *Nature,* 392, 821, 1998.
180. Felts, S. J. et al., The HSP90-related protein TRAP1 is a mitochondrial protein with distinct functional properties, *J. Biol. Chem.,* 275, 3305, 2000.
181. Michels, A. A. et al., HSP70 and HSP40 chaperone activities in the cytoplasm and the nucleus of mammalian cells, *J. Biol. Chem.,* 272, 33283, 1997.
182. Ohtsuka, K. and Hata, M., Mammalian HSP40/DNAJ homologs: cloning of novel cDNAs and a proposal for their classification and nomenclature, *Cell Stress Chaperones,* 5, 98, 2000.
183. Nollen, E. A. et al., Modulation of *in vivo* HSP70 chaperone activity by Hip and Bag-1, *J. Biol. Chem.,* 276, 4677, 2001.
184. Frydman, J. and Höhfeld, J., Chaperones get in touch: the hip-hop connection, *Trends Biochem. Sci.,* 22, 87, 1997.
185. McClellan, A. J. and Frydman, J., Molecular chaperones and the art of recognizing a lost cause, *Nat. Cell Biol.,* 3, E51, 2001.
186. Meacham, G. C. et al., The HSC70 co-chaperone CHIP targets immature CFTR for proteasomal degradation, *Nat. Cell Biol.,* 3, 100, 2001.
187. Nollen, E. A. et al., Bag1 functions *in vivo* as a negative regulator of HSP70 chaperone activity, *Mol. Cell. Biol.,* 20, 1083, 2000.
188. Takayama, S. et al., BAG-1 modulates the chaperone activity of HSP70/HSC70, *EMBO J.,* 16, 4887, 1997.
189. Song, J., Takeda, M., and Morimoto, R. I., Bag1-HSP70 mediates a physiological stress signalling pathway that regulates Raf-1/ERK and cell growth, *Nat. Cell Biol.,* 3, 276, 2001.
190. Liu, Y. and Steinacker, J. M., Changes in skeletal muscle heat shock proteins: pathological significance, *Front Biosci.,* 6, D12, 2001.
191. Su, C.-Y., Chang, C., and Lai, C.-C., Induction of heat shock proteins by exercise, in *Pharmacology in Exercise and Sports,* Somani, S. M., Ed., CRC Press, Boca Raton, FL, 1996, 147.
192. Powers, S. K., Locke, M., and Demirel, H. A., Exercise, heat shock proteins, and myocardial protection from I-R injury, *Med. Sci. Sports Exerc.,* 33, 386, 2001.
193. Nader, G. A. and Esser, K. A., Intracellular signaling specificity in skeletal muscle in response to different modes of exercise, *J. Appl. Physiol.,* 90, 1936, 2001.
194. Baar, K., Blough, E., Dineen, B., and Esser, K., Transcriptional regulation in response to exercise, *Exerc. Sport Sci. Rev.,* 27, 333, 1999.
195. Widegren, U. et al., Divergent effects of exercise on metabolic and mitogenic signaling pathways in human skeletal muscle, *FASEB J.,* 12, 1379, 1998.
196. Benjamin, I. J. et al., Induction of stress proteins in cultured myogenic cells. Molecular signals for activation of heat shock transcription factor during ischemia, *J. Clin. Invest.,* 89, 1685, 1992.
197. van Why, S. K. et al., Activation of heat-shock transcription factor by graded reductions in renal ATP, *in vivo,* in the rat, *J. Clin. Invest.,* 94, 1518, 1994.
198. Beckmann, R. P., Lovett, M., and Welch, W. J., Examining the function and regulation of HSP70 in cells subjected to metabolic stress, *J. Cell Biol.,* 117, 1137, 1992.

199. Febbraio, M. A. and Koukoulas, I., HSP72 gene expression progressively increases in human skeletal muscle during prolonged, exhaustive exercise, *J. Appl. Physiol.*, 89, 1055, 2000.

200. Mestril, R. et al., Isolation of a novel inducible rat heat-shock protein (HSP70) gene and its expression during ischaemia/hypoxia and heat shock, *Biochem. J.*, 298 (Pt. 3), 561, 1994.

201. Benjamin, I. J., Kroger, B., and Williams, R. S., Activation of the heat shock transcription factor by hypoxia in mammalian cells, *Proc. Natl. Acad. Sci. U.S.A.*, 87, 6263, 1990.

202. Cairo, G. et al., Synthesis of heat shock proteins in rat liver after ischemia and hyperthermia, *Hepatology*, 5, 357, 1985.

203. Petronini, P. G. et al., Effect of an alkaline shift on induction of the heat shock response in human fibroblasts, *J. Cell Physiol.*, 162, 322, 1995.

204. Gapen, C. J. and Moseley, P. L., Acidosis alters the hyperthermic cytotoxicity and the cellular stress response, *J. Thermal Biol.*, 20, 321, 1995.

205. Edington, B. V., Whelan, S. A., and Hightower, L. E., Inhibition of heat shock (stress) protein induction by deuterium oxide and glycerol: additional support for the abnormal protein hypothesis of induction, *J. Cell Physiol.*, 139, 219, 1989.

206. Ananthan, J., Goldberg, A. L., and Voellmy, R., Abnormal proteins serve as eukaryotic stress signals and trigger the activation of heat shock genes, *Science Wash. D.C.*, 232, 522, 1986.

207. Essig, D. A. and Nosek, T. M., Muscle fatigue and induction of stress protein genes: a dual function of reactive oxygen species?, *Can. J. Appl. Physiol.*, 22, 409, 1997.

208. Zou, J. et al., Correlation between glutathione oxidation and trimerization of heat shock factor 1, an early step in stress induction of the HSP response, *Cell Stress Chaperones*, 3, 130, 1998.

209. Salo, D. C., Donovan, C. M., and Davies, K. J., HSP70 and other possible heat shock or oxidative stress proteins are induced in skeletal muscle, heart, and liver during exercise, *Free Radic. Biol. Med.:*, 11, 239, 1991.

210. Mosser, D. D. et al., *In vitro* activation of heat shock transcription factor DNA-binding by calcium and biochemical conditions that affect protein conformation, *Proc. Natl. Acad. Sci. U.S.A.*, 87, 3748, 1990.

211. Kiang, J. G. et al., HSP-72 synthesis is promoted by increase in $[Ca^{2+}]i$ or activation of G proteins but not pHi or cAMP, *Am. J. Physiol.*, 267, C104, 1994.

212. Inaguma, Y. et al., Induction of the synthesis of HSP27 and αB-crystallin in tissues of heat-stressed rats and its suppression by ethanol or an a_1-adrenergic antagonist, *J. Biochem.*, 117, 1238, 1995.

213. Chin, J. H. et al., Activation of heat shock protein (HSP 70) and proto-oncogene expression by alpha1 adrenergic agonist in rat aorta with age, *J. Clin. Invest.*, 97, 2316, 1996.

214. Paroo, Z. and Noble, E. G., Isoproterenol potentiates exercise-induction of HSP70 in cardiac and skeletal muscle, *Cell Stress Chaperones*, 4, 199, 1999.

215. Knowlton, A. A. et al., A single myocardial stretch or decreased systolic fiber shortening stimulates the expression of heat shock protein 70 in the isolated, erythrocyte-perfused rabbit heart, *J. Clin. Invest.*, 88, 2018, 1991.

216. Kilgore, J. L. et al., Stress protein induction in skeletal muscle: comparison of laboratory models to naturally occurring hypertrophy, *J. Appl. Physiol.*, 76, 598, 1994.

217. Thompson, H. S. et al., A single bout of eccentric exercise increases HSP27 and HSC/HSP70 in human skeletal muscle, *Acta Physiol. Scand.*, 171, 187, 2001.
218. Moseley, P. L., Heat shock proteins and the inflammatory response, *Ann. N.Y. Acad. Sci.*, 856, 206, 1998.
219. Jäättelä, M. and Wissing, D., Emerging role of heat shock proteins in biology and medicine, *Ann. Med.*, 24, 249, 1992.
220. Hammond, G. L., Lai, Y. K., and Markert, C. L., Diverse forms of stress lead to new patterns of gene expression through a common and essential metabolic pathway, *Proc. Natl. Acad. Sci. U.S.A.*, 79, 3485, 1982.
221. Paroo, Z., Tiidus, P. M., and Noble, E. G., Estrogen attenuates HSP72 expression in acutely exercised male rodents, *Eur. J. Appl. Physiol.*, 80, 180, 1999.
222. Locke, M., Noble, E. G., and Atkinson, B. G., Exercising mammals synthesize stress proteins, *Am. J. Physiol.*, 258, C723, 1990.
223. Hernando, R. and Manso, R., Muscle fibre stress in response to exercise: synthesis, accumulation and isoform transitions of 70-kDa heat-shock proteins, *Eur. J. Biochem.*, 243, 460, 1997.
224. Skidmore, R. et al., HSP70 induction during exercise and heat stress in rats: role of internal temperature, *Am. J. Physiol.*, 268, R92, 1995.
225. Taylor, R. P., Harris, M. B., and Starnes, J. W., Acute exercise can improve cardioprotection without increasing heat shock protein content, *Am. J. Physiol.*, 276, H1098, 1999.
226. Walters, T. J. et al., HSP70 expression in the CNS in response to exercise and heat stress in rats, *J. Appl. Physiol.*, 84, 1269, 1998.
227. Kregel, K. C. and Moseley, P. L., Differential effects of exercise and heat stress on liver HSP70 accumulation with aging, *J. Appl. Physiol.*, 80, 547, 1996.
228. Locke, M. et al., Enhanced postischemic myocardial recovery following exercise induction of HSP72, *Am. J. Physiol.*, 269, H320, 1995.
229. Locke, M. et al., Activation of heat-shock transcription factor in rat heart after heat shock and exercise, *Am. J. Physiol.*, 268, C1387, 1995.
230. Smolka, M. B. et al., HSP72 as a complementary protection against oxidative stress induced by exercise in the soleus muscle of rats, *Am. J. Physiol. Regul. Integr. Comp. Physiol.*, 279, R1539, 2000.
231. McArdle, A. et al., Contractile activity-induced oxidative stress: cellular origin and adaptive responses, *Am. J. Physiol. Cell Physiol.*, 280, C621, 2001.
232. Fehrenbach, E. et al., Transcriptional and translational regulation of heat shock proteins in leukocytes of endurance runners, *J. Appl. Physiol.*, 89, 704, 2000.
233. Fehrenbach, E. et al., HSP expression in human leukocytes is modulated by endurance exercise, *Med. Sci. Sports Exerc.*, 32, 592, 2000.
234. Puntschart, A. et al., HSP70 expression in human skeletal muscle after exercise, *Acta Physiol. Scand.*, 157, 411, 1996.
235. Reichsman, F. et al., Muscle protein changes following eccentric exercise in humans, *Eur. J. Appl. Physiol.*, 62, 245, 1991.
236. Pilegaard, H. et al., Transcriptional regulation of gene expression in human skeletal muscle during recovery from exercise, *Am. J. Physiol. Endocrinol. Metab.*, 279, E806, 2000.
237. Khassaf, M. et al., Time course of responses of human skeletal muscle to oxidative stress induced by nondamaging exercise, *J. Appl. Physiol.*, 90, 1031, 2001.
238. Thompson, H. S. and Scordilis, S. P., Ubiquitin changes in human biceps muscle following exercise-induced damage, *Biochem. Biophys. Res. Commun.*, 204, 1193, 1994.

239. Trudell, J. R. et al., Induction of HSP72 in rat liver by chronic ethanol consumption combined with exercise: association with the prevention of ethanol-induced fatty liver by exercise, *Alcohol Clin. Exp. Res.,* 19, 753, 1995.

240. Muller, E. et al., Effects of long-term changes in medullary osmolality on heat shock proteins HSP25, HSP60, HSP72 and HSP73 in the rat kidney, *Pflugers Arch.,* 435, 705, 1998.

241. Sorger, P. K. and Pelham, H. R., Cloning and expression of a gene encoding HSC73, the major HSP70-like protein in unstressed rat cells, *EMBO J.,* 6, 993, 1987.

242. Ryan, A. J., Gisolfi, C. V., and Moseley, P. L., Synthesis of 70K stress protein by human leukocytes: effect of exercise in the heat, *J. Appl. Physiol.,* 70, 466, 1991.

243. Gonzalez, B., Hernando, R., and Manso, R., Stress proteins of 70 kDa in chronically exercised skeletal muscle, *Pflugers Arch.,* 440, 42, 2000.

244. Demirel, H. A. et al., The effects of exercise duration on adrenal HSP72/73 induction in rats, *Acta Physiol. Scand.,* 167, 227, 1999.

245. Liu, Y. et al., Human skeletal muscle HSP70 response to training in highly trained rowers, *J. Appl. Physiol.,* 86, 101, 1999.

246. Neufer, P. D. et al., Continuous contractile activity induces fiber type specific expression of HSP70 in skeletal muscle, *Am. J. Physiol.,* 271, C1828, 1996.

247. Noble, E. G. et al., Differential expression of stress proteins in rat myocardium after free wheel or treadmill run training, *J. Appl. Physiol.,* 86, 1696, 1999.

248. Liu, Y. et al., Human skeletal muscle HSP70 response to physical training depends on exercise intensity, *Int. J. Sports Med.,* 21, 351, 2000.

249. Dudley, G. A., Abraham, W. M., and Terjung, R. L., Influence of exercise intensity and duration on biochemical adaptations in skeletal muscle, *J. Appl. Physiol.,* 53, 844, 1982.

250. Friden, J. and Lieber, R. L., Eccentric exercise-induced injuries to contractile and cytoskeletal muscle fibre components, *Acta Physiol. Scand.,* 171, 321, 2001.

251. Ingalls, C. P., Warren, G. L., and Armstrong, R. B., Dissociation of force production from MHC and actin contents in muscle injured by eccentric contractions, *J. Muscle Res. Cell Motil.,* 19, 215, 1998.

252. Sonneborn, J. S. and Barbee, S. A., Exercise-induced stress response as an adaptive tolerance strategy, *Environ. Health Perspect.,* 106 (Suppl. 1), 325, 1998.

253. Roussel, D. et al., Differential effects of endurance training and creatine depletion on regional mitochondrial adaptations in rat skeletal muscle, *Biochem. J.,* 350, 547, 2000.

254. Mattson, J. P. et al., Induction of mitochondrial stress proteins following treadmill running, *Med. Sci. Sports Exerc.,* 32, 365, 2000.

255. Samelman, T. R., Heat shock protein expression is increased in cardiac and skeletal muscles of Fischer 344 rats after endurance training, *Exp. Physiol.,* 85, 92, 2000.

256. Powers, S. K. et al., Exercise training improves myocardial tolerance to *in vivo* ischemia-reperfusion in the rat, *Am. J. Physiol.,* 275, R1468, 1998.

257. Demirel, H. A. et al., Exercise training reduces myocardial lipid peroxidation following short-term ischemia-reperfusion, *Med. Sci. Sports Exerc.,* 30, 1211, 1998.

258. Samelman, T. R., Shiry, L. J., and Cameron, D. F., Endurance training increases the expression of mitochondrial and nuclear encoded cytochrome c oxidase subunits and heat shock proteins in rat skeletal muscle, *Eur. J. Appl. Physiol.,* 83, 22, 2000.

259. Gonzalez, B., Hernando, R., and Manso, R., Anabolic steroid and gender-dependent modulation of cytosolic HSP70s in fast- and slow-twitch skeletal muscle, *J. Steroid Biochem. Mol. Biol.*, 74, 63, 2000.

260. Naito, H. et al., Exercise training increases heat shock protein in skeletal muscles of old rats, *Med. Sci. Sports Exerc.*, 33, 729, 2001.

261. Harris, M. B. and Starnes, J. W., Effects of body temperature during exercise training on myocardial adaptations, *Am. J. Physiol. Heart Circ. Physiol.*, 280, H2271, 2001.

262. Ecochard, L. et al., Skeletal muscle HSP72 level during endurance training: influence of peripheral arterial insufficiency, *Pflugers Arch.*, 440, 918, 2000.

263. Locke, M. and Tanguay, R. M., Increased HSF activation in muscles with a high constitutive HSP70 expression, *Cell Stress Chaperones*, 1, 189, 1996.

264. Boshoff, T. et al., Differential basal synthesis of HSP70/HSC70 contributes to interindividual variation in HSP70/HSC70 inducibility, *Cell Mol. Life Sci.*, 57, 1317, 2000.

265. Lyashko, V. N. et al., Comparison of the heat shock response in ethnically and ecologically different human populations, *Proc. Natl. Acad. Sci. U.S.A.*, 91, 12492, 1994.

266. Locke, M. and Tanguay, R. M., Diminished heat shock response in the aged myocardium, *Cell Stress Chaperones*, 1, 251, 1996.

267. DiDomenico, B. J., Bugaisky, G. E., and Lindquist, S., The heat shock response is self-regulated at both the transcriptional and posttranscriptional levels, *Cell*, 31, 593, 1982.

268. Neufer, P. D. and Benjamin, I. J., Differential expression of αB-crystallin and HSP27 in skeletal muscle during continuous contractile activity, *J. Biol. Chem.*, 271, 24089, 1996.

269. Ornatsky, O. I., Conner, M. K., and Hood, D. A., Expression of stress proteins and mitochondrial chaperonins in chronically stimulated skeletal muscle, *Biochem. J.*, 311, 119, 1995.

270. Takahashi, M. et al., Contractile activity-induced adaptations in the mitochondrial protein import system, *Am. J. Physiol.*, 274, C1380, 1998.

271. Takahashi, M. and Hood, D. A., Protein import into subsarcolemmal and inter-myofibrillar skeletal muscle mitochondria. Differential import regulation in distinct subcellular regions, *J. Biol. Chem.*, 271, 27285, 1996.

272. Pette, D. and Vrbova, G., What does chronic electrical stimulation teach us about muscle plasticity?, *Muscle Nerve*, 22, 666, 1999.

273. Sultan, K. R. et al., Fiber type-specific expression of major proteolytic systems in fast- to slow-transforming rabbit muscle, *Am. J. Physiol. Cell Physiol.*, 280, C239, 2001.

274. Medina, R., Wing, S. S., and Goldberg, A. L., Increase in levels of polyubiquitin and proteasome mRNA in skeletal muscle during starvation and denervation atrophy, *Biochem. J.*, 307, 631, 1995.

275. Wing, S. S., Hass, A. L., and Goldberg, A. L., Increase in ubiquitin-protein conjugates concomitant with the increase in proteolysis in rat skeletal muscle during starvation and atrophy denervation, *Biochem. J.*, 307, 639, 1995.

276. Haas, A. L. and Riley, D. A., The dynamics of ubiquitin pools within skeletal muscle, in *The Ubiquitin System*, Schlesinger, M. and Hershko, A., Eds., Cold Spring Harbor Laboratory Press, Cold Spring Harbor, NY, 1988, 178.

277. Riley, D. A. et al., Quantification and immunocytochemical localization of ubiquitin conjugates within rat red and white skeletal muscles, *J. Histochem. Cytochem.*, 36, 621, 1988.

278. Tews, D. S. et al., Expression profile of stress proteins, intermediate filaments, and adhesion molecules in experimentally denervated and reinnervated rat facial muscle, *Expt. Neurol.*, 146, 125, 1997.

279. Locke, M., Noble, E. G., and Atkinson, B. G., Inducible isoform of HSP70 is constitutively expressed in a muscle fiber type specific pattern, *Am. J. Physiol.*, 261, C774, 1991.

280. Vesely, M. J. et al., Fibre type specificity of haem oxygenase-1 induction in rat skeletal muscle, *FEBS Lett.*, 458, 257, 1999.

281. McGrath, L. B. et al., Heat shock protein (HSP72) expression in patients undergoing cardiac surgery, *J. Thorac. Cardiovasc. Surg.*, 109, 370, 1995.

282. Knowlton, A. A. et al., Differential expression of heat shock proteins in normal and failing human hearts, *J. Mol. Cell Cardiol.*, 30, 811, 1998.

283. Jenkins, R. R. and Goldfarb, A., Introduction: oxidant stress, aging, and exercise, *Med. Sci. Sports Exerc.*, 25, 210, 1993.

284. Freeman, M. L. et al., On the path to the heat shock response: destabilization and formation of partially folded protein intermediates, a consequence of protein thiol modification, *Free Radic. Biol. Med.*, 26, 737, 1999.

285. Naito, H. et al., Heat stress attenuates skeletal muscle atrophy in hindlimb-unweighted rats, *J. Appl. Physiol.*, 88, 359, 2000.

286. Ku, Z. et al., Decreased polysomal HSP-70 may slow polypeptide elongation during skeletal muscle atrophy, *Am. J. Physiol.*, 268, C1369, 1995.

287. Atomi, Y., Yamada, S., and Nishida, T., Early changes of αB-crystallin mRNA in rat skeletal muscle to mechanical tension and denervation, *Biochem. Biophys. Res. Commun.*, 181, 1323, 1991.

288. Buonanno, A. and Rosenthal, N., Molecular control of muscle diversity and plasticity, *Dev. Genet.*, 19, 95, 1996.

289. Gopal-Srivastava, R. and Piatigorsky, J., The murine αB-crystallin/small heat shock protein enhancer: identification of αBE-1, αBE-2, αBE-3, and MRF control elements, *Mol. Cell. Biol.*, 13, 7144, 1993.

290. Gopal-Srivastava, R., Haynes, J. I. I., and Piatigorsky, J., Regulation of the murine αB-crystallin/small heat shock protein gene in cardiac muscle, *Mol. Cell. Biol.*, 15, 7081, 1995.

291. Noble, E. G. and Aubrey, F. K., Stress proteins and the adaptive response to exercise and muscle loading, *Biochem Exercise*, 10, 329, 1998.

292. Willoughby, D. S., Priest, J. W., and Jennings, R. A., Myosin heavy chain isoform and ubiquitin protease mRNA expression after passive leg cycling in persons with spinal cord injury, *Arch. Phys. Med. Rehabil.*, 81, 157, 2000.

293. Marber, M. S. et al., Overexpression of the rat inducible 70-kD heat stress protein in a transgenic mouse increases the resistance of the heart to ischemic injury, *J. Clin. Invest.*, 95, 1446, 1995.

294. Trost, S. U. et al., Protection against myocardial dysfunction after a brief ischemic period in transgenic mice expressing inducible heat shock protein, *J. Clin. Invest.*, 101, 855, 1998.

295. Plumier, J.-C. L. et al., Transgenic mice expressing the human heat shock protein 70 have improved post-ischemic myocardial recovery, *J. Clin. Invest.*, 95, 1854, 1995.

296. Radford, N. B. et al., Cardioprotective effects of 70-kDa heat shock protein in transgenic mice, *Proc. Natl. Acad. Sci. U.S.A.*, 93, 2339, 1996.

297. Suzuki, K. et al., Reduction in myocardial apoptosis associated with overexpression of heat shock protein 70, *Basic Res. Cardiol.*, 95, 397, 2000.

298. Jayakumar, J. et al., Gene therapy for myocardial protection: transfection of donor hearts with heat shock protein 70 gene protects cardiac function against ischemia-reperfusion injury, *Circulation*, 102, III302, 2000.

299. Suzuki, K. et al., *In vivo* gene transfection with heat shock protein 70 enhances myocardial tolerance to ischemia-reperfusion injury in rat, *J. Clin. Invest.*, 99, 1645, 1997.

300. Donnelly, T. J. et al., Heat shock protein induction in rat hearts. A role for improved myocardial salvage after ischemia and reperfusion?, *Circulation*, 85, 769, 1992.

301. Kukreja, R. C. et al., Role of protein kinase C and 72 kDa heat shock protein in ischemic tolerance following heat stress in the rat heart, *Mol. Cell. Biochem.*, 195, 123, 1999.

302. Morris, J. N. et al., Vigorous exercise in leisure-time: protection against coronary heart disease., *The Lancet*, 2, 1207, 1980.

303. Barinaga, M., How much pain for cardiac gain?, *Science*, 276, 1324, 1997.

304. Garramone, R. R., Jr. et al., Reduction of skeletal muscle injury through stress conditioning using the heat-shock response, *Plast. Reconstr. Surg.*, 93, 1242, 1994.

305. Lepore, D. A. et al., Prior heat stress improves survival of ischemic-reperfused skeletal muscle *in vivo*, *Muscle Nerve*, 23, 1847, 2000.

306. Wang, B. H. et al., Improved free musculocutaneous flap survival with induction of heat shock protein, *Plast. Reconstr. Surg.*, 101, 776, 1998.

307. Lille, S. et al., Induction of heat-shock protein 72 in rat skeletal muscle does not increase tolerance to ischemia-reperfusion injury, *Muscle Nerve*, 22, 390, 1999.

308. Lepore, D. A. and Morrison, W. A., Ischemic preconditioning: lack of delayed protection against skeletal muscle ischemia-reperfusion, *Microsurgery*, 20, 350, 2000.

309. King, D. W. and Gollnick, P. D., Ultrastructure of rat heart and liver after exhaustive exercise, *Am. J. Physiol.*, 218, 1150, 1970.

310. Phaneuf, S. and Leeuwenburgh, C., Apoptosis and exercise, *Med. Sci. Sports Exerc.*, 33, 393, 2001.

311. Finkel, E., The mitochondrion: is it central to apoptosis?, *Science*, 292, 624, 2001.

312. Boppart, M. D. et al., Marathon running transiently increases c-Jun NH2-terminal kinase and p38 activities in human skeletal muscle, *J. Physiol.*, 526 (Pt. 3), 663, 2000.

313. Goodyear, L. J. et al., Effects of exercise and insulin on mitogen-activated protein kinase signaling pathways in rat skeletal muscle, *Am. J. Physiol.*, 271, E403, 1996.

314. Boppart, M. D. et al., Eccentric exercise markedly increases c-Jun NH(2)-terminal kinase activity in human skeletal muscle, *J. Appl. Physiol.*, 87, 1668, 1999.

315. Armstrong, R. B., Warren, G. L., and Warren, J. A., Mechanisms of exercise-induced muscle fibre injury, *Sports Med.*, 12, 184, 1991.

316. Meriin, A. B. et al., Protein-damaging stresses activate c-Jun N-terminal kinase via inhibition of its dephosphorylation: a novel pathway controlled by HSP72 [published erratum appears in *Mol. Cell. Biol.*, 19, 5235, 1999], *Mol. Cell. Biol.*, 19, 2547, 1999.

317. Beere, H. M. et al., Heat-shock protein 70 inhibits apoptosis by preventing recruitment of procaspase-9 to the Apaf-1 apoptosome, *Nat. Cell Biol.*, 2, 469, 2000.

318. Saleh, A. et al., Negative regulation of the Apaf-1 apoptosome by HSP70, *Nat. Cell Biol.*, 2, 476, 2000.

319. Samali, A. et al., Presence of a pre-apoptotic complex of pro-caspase-3, HSP60 and HSP10 in the mitochondrial fraction of jurkat cells, *EMBO J.*, 18, 2040, 1999.

320. Xanthoudakis, S. et al., HSP60 accelerates the maturation of pro-caspase-3 by upstream activator proteases during apoptosis, *EMBO J.*, 18, 2049, 1999.
321. Ray, P. S. et al., Transgene overexpression of αB-crystallin confers simultaneous protection against cardiomyocyte apoptosis and necrosis during myocardial ischemia and reperfusion, *FASEB J.*, 15, 393, 2001.
322. Barclay, J. K. and Hansel, M., Free radicals may contribute to oxidative skeletal muscle fatigue, *Can. J. Physiol. Pharmacol.*, 69, 279, 1991.
323. Brotto, M. A. and Nosek, T. M., Hydrogen peroxide disrupts Ca^{2+} release from the sarcoplasmic reticulum of rat skeletal muscle fibers, *J. Appl. Physiol.*, 81, 731, 1996.
324. Thomas, J. A. and Noble, E. G., Heat shock does not attenuate low-frequency fatigue, *Can. J. Physiol. Pharmacol.*, 77, 64, 1999.
325. Chin, E. R. and Allen, D. G., The role of elevations in intracellular $[Ca^{2+}]$ in the development of low frequency fatigue in mouse single muscle fibres, *J. Physiol.*, 491 (Pt. 3), 813, 1996.
326. Nosek, T. M. et al., Functional properties of skeletal muscle from transgenic animals with upregulated heat shock protein 70, *Physiol. Genomics*, 4, 25, 2000.
327. Arispe, N. and De Maio, A., ATP and ADP modulate a cation channel formed by HSC70 in acidic phospholipid membranes, *J. Biol. Chem.*, 275, 30839, 2000.
328. Thomason, D. B., Anderson, O., III, and Menon, V., Fractal analysis of cytoskeleton rearrangement in cardiac muscle during head-down tilt, *J. Appl. Physiol.*, 81, 1522, 1996.
329. Rennie, M. J. and Tipton, K. D., Protein and amino acid metabolism during and after exercise and the effects of nutrition, *Annu. Rev. Nutr.*, 20, 457, 2000.
330. Someren, J. S. et al., Heat shock proteins 70 and 90 increase calcineurin activity *in vitro* through calmodulin-dependent and independent mechanisms, *Biochem. Biophys. Res. Commun.*, 260, 619, 1999.
331. Tumlin, J. A. et al., Aldosterone and dexamethasone stimulate calcineurin activity through a transcription-independent mechanism involving steroid receptor-associated heat shock proteins, *J. Clin. Invest.*, 99, 1217, 1997.
332. Chin, E. R. et al., A calcineurin-dependent transcriptional pathway controls skeletal muscle fiber type, *Genes Dev.*, 12, 2499, 1998.
333. Bigard, X. et al., Calcineurin co-regulates contractile and metabolic components of slow muscle phenotype, *J. Biol. Chem.*, 275, 19653, 2000.
334. Dunn, S. E., Burns, J. L., and Michel, R. N., Calcineurin is required for skeletal muscle hypertrophy, *J. Biol. Chem.*, 274, 21908, 1999.
335. Dunn, S. E., Chin, E. R., and Michel, R. N., Matching of calcineurin activity to upstream effectors is critical for skeletal muscle fiber growth, *J. Cell Biol.*, 151, 663, 2000.
336. Moseley, P. L., Exercise, stress, and the immune conversation, *Exerc. Sport Sci. Rev.*, 28, 128, 2000.
337. Folch, E. et al., Pancreatitis induces HSP72 in the lung: role of neutrophils and xanthine oxidase, *Biochem. Biophys. Res. Commun.*, 273, 1078, 2000.
338. Booth, F. W. et al., Waging war on modern chronic diseases: primary prevention through exercise biology, *J. Appl. Physiol.*, 88, 774, 2000.
339. Krook, A. et al., Effects of exercise on mitogen- and stress-activated kinase signal transduction in human skeletal muscle, *Am. J. Physiol. Regul. Integr. Comp. Physiol.*, 279, R1716, 2000.
340. Sheikh-Hamad, D. et al., p38 kinase activity is essential for osmotic induction of mRNAs for HSP70 and transporter for organic solute betaine in Madin-Darby canine kidney cells, *J. Biol. Chem.*, 273, 1832, 1998.

341. Mizukoshi, E. et al., Fibroblast growth factor-1 interacts with the glucose-regulated protein GRP75/mortalin, *Biochem. J.*, 343 (Pt. 2), 461, 1999.
342. Kim, J. et al., Analysis of the phosphorylation of human heat shock transcription factor-1 by MAP kinase family members, *J. Cell Biochem.*, 67, 43, 1997.
343. Tidball, J. G. et al., Mechanical loading regulates NOS expression and activity in developing and adult skeletal muscle, *Am. J. Physiol.*, 275, C260, 1998.
344. Dunbar, C. C. and Kalinski, M. I., Cardiac intracellular regulation: exercise effects on the cAMP system and A-kinase, *Med. Sci. Sports Exerc.*, 26, 1459, 1994.
345. Choi, H. S. et al., cAMP and cAMP-dependent protein kinase regulate the human heat shock protein 70 gene promoter activity, *J. Biol. Chem.*, 266, 11858, 1991.
346. Malyshev, I. Yu. et al., Nitric oxide is involved in heat-induced HSP70 accumulation, *FEBS Lett.*, 370, 159, 1995.
347. Ding, X. Z., Tsokos, G. C., and Kiang, J. G., Overexpression of HSP-70 inhibits the phosphorylation of HSF1 by activating protein phosphatase and inhibiting protein kinase C activity, *FASEB J.*, 12, 451, 1998.

chapter four

Hsps and protein synthesis in striated muscle

Donald B. Thomason and Vandana Menon

Contents

I. Introduction

Lying in the balance between the anabolic process of protein synthesis and the catabolic process of protein degradation is protein expression. For physiologic viability, protein expression must be controlled, thus requiring control of both protein synthesis and degradation. This chapter reviews the role of heat shock proteins 70 and 25 in the control of proteins synthesis.

Eukaryotic protein synthesis consists of three general phases, each with their own points of control:[1] initiation of the protein synthesis complex, elongation of the nascent polypeptide, and termination by release of the nascent protein and disassembly of the synthesis complex. Initiation of protein synthesis involves a myriad of cofactors to recognize mRNA, assemble the ribosome, and locate the initiation site.[1-4] As discussed below, heat shock proteins (HSPs) can affect initiation. Elongation of the nascent polypeptide involves coordinate movement of the ribosome, deciphering of the next codon, and condensation of a new amino acid on the growing protein.[5] Heat shock proteins can affect elongation as well. Termination of synthesis and release of the new protein has received less research attention, but is no less important for protein expression.[6] Although lacking direct evidence, the apparent influence of C-terminal structure on termination rate[7] leaves open the possibility for influence by molecular chaperones. As will become obvious in the discussion that follows, heat shock proteins have the potential for many interactions that may affect protein synthesis in mammalian striated muscle.

II. Heat shock protein 70 and protein synthesis

The heat shock protein 70 (HSP70) family of molecular chaperones has been the focus of intense interest during the past decade. It is now well-recognized that these proteins and their associated proteins provide dual functionality. On the one hand, they protect the cell from the aberrantly folded proteins as may occur during cell stress, hence their initial characterization as "heat shock" or stress proteins. On the other hand, these molecular chaperones aid in the normal, unstressed transport and folding of nascent polypeptides and in the refolding of denatured proteins. These functions are not exclusive and, true to the principles of homeostasis, the molecular chaperones act seamlessly to provide the actions necessary for cells to meet the challenges they encounter. The section that follows discusses these actions in mammalian skeletal muscle, particularly as they relate to protein synthesis.

A.   HSP70 function

1.   Binding to denatured protein

As with several cell types, some skeletal muscle fibers constitutively express a cognate and an inducible heat shock protein 70 (HSC70 and HSP70, respectively), in addition to having the classic ability to inducibly express the HSP70.[8-12] To briefly review, the HSP/HSC70 proteins are thought to aid in the proper folding or refolding of proteins.[13-16] As a stress response, expression of HSP70 provides clear survival advantages to the cell.[17-19] The HSP70 family of proteins are nucleotide binding proteins, the nucleotide affecting the affinity of the chaperone for unfolded protein: ADP-bound HSP70 has a higher affinity for unfolded protein than does the ATP-bound form.[20-23] The endogenous ATPase activity of HSP70[24,25] is affected by various cofactors (or co-chaperones), thus affecting the affinity for unfolded protein. Specifically, the HSP40 class of proteins significantly increases HSP70 ATPase activity, resulting in an increase in affinity of the HSP70-HSP40 complex for the unfolded protein.[26,27] Members of the HSP110 protein family also have chaperone activity.[28] Pertinent to this discussion, specific HSP110 class members (HSP105α,β) can modulate the HSP70-HSP40 ATPase activity by increasing the rate of hydrolysis.[29] In turn, the increase in ATPase activity increases the affinity of the complex for unfolded protein. A counter-regulatory mechanism also exists in the form of an HSP70 binding protein, HSP70BP1. Found abundantly in cardiac and skeletal muscle, HSP70BP1 significantly inhibits the HSP40-stimulated ATPase activity of HSP70.[30,31] Subsequent to binding of the denatured protein by HSP70, HSP40, HSP110, and apparently HSP25,[32] renaturation or degradation takes place through the action of HSP90 and other accessory proteins.[15,33,34] The regulation of the unfolded protein binding is obviously complex, considering the interaction and potential modulatory effects of the various components within the binding complex. It is therefore worthwhile to consider the unstressed physiological function without compounding the complexity with the induction of the proteins by stress. The following paragraphs briefly discuss the normal functioning of the binding complex.

In the cytosol, the HSP70 class of proteins can interact with the ribonucleoprotein complex that consists of the ribosome, mRNA, and nascent polypeptide.[1,9,14,35] Although the HSP70 is associated with the polyribosomes (a.k.a., polysomes),[9,10] the most direct evidence for an interaction with the nascent polypeptides comes from puromycin-truncated polypeptides. Puromycin causes premature termination of translation and release of the truncated nascent polypeptide.[36] In yeast, truncation with puromycin disrupts the association of HSP70 with the polysomes,[35] and immunoprecipitation with an antibody to puromycin co-precipitates predominantly HSP70.[37] Others and we have shown that this association of HSP70 with the nascent polypeptide is acutely sensitive to cytosolic ATP levels (Figure 4.1A).[1,9,10,14,37] The coordinate assembly of nascent polypeptide, HSP70, and HSP40 apparently is an essential step toward the productive folding of the newly synthesized protein into a functional protein, and differs from the aggregation of heat shock proteins on denatured proteins.[38,39] Subsequent productive folding of the nascent polypeptide then requires participation of members of the chaperonin or HSP60 family of proteins. The association of HSP70 and HSP40 with the growing polypeptide appears to be essential not only for proper folding, but also for a normal pace of protein synthesis. For example, without an HSP40 family member, full-length inactive rhodanese can be synthesized but only attains an active conformation with HSP40 addition.[40-42] Folding can apparently begin prior to completion of a full-length polypeptide, as evidenced by the appearance of partially folded intermediates.[40,43] Either this folding or the direct interaction with molecular chaperones assists in the movement of the nascent polypeptide through the ribosome channel. Temperature-sensitive yeast mutants that lack functional HSP70 activity appear to exhibit slowed nascent polypeptide elongation.[13,35,44] Overexpression of an elongation factor 1α-like protein overcomes the deficit in elongation.[35] Supporting the role of HSP70 in efficient translation, mutation of the yeast HSP40

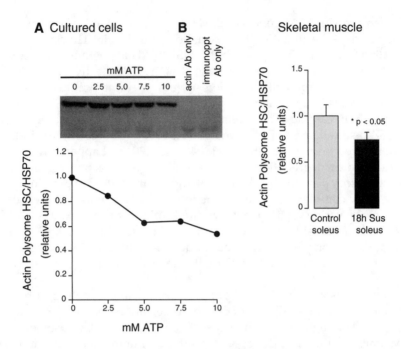

Figure 4.1   Skeletal α-actin synthesizing polysomes from (A) cultured myotubes and (B) rat soleus muscle were precipitated with a monoclonal antibody to the N-terminus of actin and probed for HSC/HSP70 content by Western blot. (A) Intracellular ATP levels affect the association of HSC/HSP70 with the actin synthesizing polysomes. ATP levels were adjusted by permeabilizing the myotubes in the presence of different extracellular concentrations of ATP.[9] Increased ATP levels decrease the protein synthesis rate.[9] (B) Soleus muscle α-actin synthesizing polyribosomes show decreased association of HSC/HSP70 with the nascent polypeptide 18 hours following hind-limb suspension, consistent with the decrease in nascent polypeptide elongation rate.[9,46]

analog results in less efficient expression of foreign genes in yeast.[45] Consistent with these data, decreased association of HSC/HSP70 with nascent polypeptides in slow-twitch skeletal muscle is associated with slowed polypeptide elongation.[9,46] As with cytosolic proteins, proteins translocated across membranes also appear to require a "tugging" action for efficient synthesis. For example, members of the HSP70 family are necessary for translocation of protein across the mitochondrial membrane.[47–49] Similarly, transport across the endoplasmic reticulum membrane also requires HSP70 family proteins.[50,51]

Indirectly, HSP70 can also affect initiation of translation. The hydrolysis of GTP and release of initiation factor 2α (eIF-2α) in a GDP-bound form is required for the formation of the 80S ribosome.[1] For a new round of initiation to take place, nucleotide exchange must take place to regenerate eIF-2α-GTP. Phosphorylation of eIF-2α by a specific kinase, eIF-2α kinase, effectively locks the eIF-2α in the GDP-bound form and inhibits a new round of initiation.[52] Normally, the eIF-2α kinase is kept in an inactive state by HSP70.[53–55] The appearance of unfolded protein, however, causes the translocation of HSP70 from the kinase to the unfolded protein. This in turn activates the kinase and slows protein synthesis initiation both *in vitro* and *in vivo*.[10,54–56]

2. Novel functions

In addition to the well-accepted chaperone function of HSP70 for binding nascent or denatured protein, reports are beginning to appear indicating novel functions for HSP70. Someren et al.[57] report a modulatory role for HSP70 on calcineurin (protein phosphatase 3) enzyme activity. In these experiments, a calmodulin-dependent increase in the V_{max} of calcineurin (protein phosphatase 3) by HSP70 and HSP90 was observed. In addition to binding protein, recent reports show an association of HSP70 and HSP110 with RNA.[58–60] The fact that some of these associations involve mRNAs is suggestive of an involvement in RNA folding in such a manner as to affect mRNA stability or expression. The binding of HSP70 and HSP110 to AU-rich regions of mRNA raises the possibility that these chaperones affect mRNA half-life.[58] HSP70 is required for the binding of erythropoietin mRNA binding protein to the 3'-untranslated region of erythropoietin mRNA, further suggesting a role of the chaperone in mRNA stability.[59]

A much less direct, although considerably more diverse, effect of HSP70 on cell function is indicated from data on the signal pathways involved with HSP70 mRNA induction. In vascular smooth muscle, stress proteins are induced by hypertensive or hypertrophic stressors.[61–63] The cyclic stretch of the muscle cells in culture results in HSP70 induction through intracellular signaling pathways that involve the small G-proteins Rac and Ras.[64,65] Although in these cells the p38[MAPK] pathway does not appear to be involved,[64] this is not true for the osmotic stress induction of HSP70 in renal cells.[66] What is interesting in both of these processes is the potential role of HSP70 as a negative feedback signal to attenuate the initial signal. In the osmotically challenged renal cells, at least initially, p38[MAPK] is stimulatory for its own expression as well as HSP70 expression,[66] and therefore must have a negative feedback mechanism to turn off the signal. Likewise, cyclic stretch creates a continued activation of the small G-proteins Rac and Ras following 6 hours of stress despite the beginning of a diminution of HSF1 binding activity and the beginning of down-regulation of HSP70 protein expression.[64] Clearly, considerable work remains to dissect these pathways.

B. HSP70, exercise, and protein synthesis

An acute bout of exercise is a potent modulator of protein synthesis, both during and following the exercise. For many proteins, the modulation of translation on such a rapid time scale is inconsistent with the time necessary for transcriptional regulation to affect translation to the degree that protein synthesis is affected. For example, a single bout of running or swimming exercise by untrained rats immediately decreases protein synthesis in the hindlimb muscles by 35 to 80%, depending on the method of measurement.[67–71] Acute resistance exercise does not depress post-exercise protein synthesis rates to the same extent.[70,72,73] The depression of protein synthesis is modulated by contractile state, correlates with the energy status of the working muscle, and appears to be inhibited at the level of initiation.[71,74]

Part of the rapid regulation of protein synthesis appears to be mediated by the HSP70 class of stress proteins. The stress of exercise induces the expression of HSP70 in a dose-dependent manner.[75–82] Because exercise or oxidative stress causes muscle damage and free radical formation,[83–86] both inducible and cognate forms of HSP70 shift toward affected proteins and away from their "normal" chaperone roles. The shift of HSP70 among intracellular pools has a significant influence on protein synthesis, either directly or through cofactors.[9,10,46,53–56] As shown in Figure 4.1, HSP70 shifts away from nascent polypeptide, an effect that decreases the nascent polypeptide elongation rate.[9,46] It is possible that a constitutive level of inducible HSP70 expression in slow-twitch skeletal muscle fibers[8,11] may have a sparing effect on protein synthesis by buffering these translocations of the chaperone protein in response to an acute bout of exercise. Perhaps a similar buffering role occurs with exercise training. Exercise training results in greater levels of expression not only in working muscle, but other tissues as well.[77,78,87–91] As shown in Figure 4.2, the acute induction of HSP70 with a single bout of exercise does apparently buffer the translocation in some, but not all muscles.

Thus, we see that HSP70, whether constitutively or inducibly expressed, can influence protein synthesis by its movement among intracellular pools. Moreover, exercise and exercise training may have profound effects not only on the expression of the chaperone, but also on the control of protein synthesis that is a consequence of the chaperone expression.

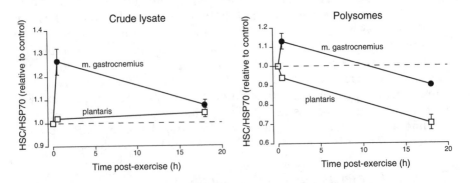

Figure 4.2   An acute bout of running exercise by rats (30 min, 20 m/min, 30% grade) induces expression of HSC/HSP70 in the medial gastrocnemius and plantaris muscles (left). Significant induction is detectable immediately following exercise. However, the HSC/HSP70 associated with polysomes isolated from the plantaris muscle indicate decreased association with the nascent polypeptides (right). This is consistent with the translocation of the stress protein between intracellular pools.

III. Heat shock protein 25 and protein synthesis

The heat shock protein 25 (HSP25) class of stress proteins (including murine HSP25 and human HSP27) contains at least three family members, and likely more.[92] A common naming scheme has been adopted for the family member genes, HSPB,[92] of which the most generically referred to member that is expressed is from the gene HSPB1. HSP25 bears sequence homology with the HSP20 family of small heat shock proteins, of which αB-crystallin is perhaps the best-known member.[93-95] Although αB-crystallin is known for its expression in the lens of the eye, it is also expressed in other tissues. In particular, αB-crystallin expression is greater in slow-twitch than fast-twitch skeletal muscle, and is regulated in response to mechanical activity and thyroid status.[96,97] This section addresses the functions of HSP25 and similar proteins, with particular attention given to protein synthesis and exercise.

A. HSP25 function

1. Binding to denatured protein and cellular dynamics

The precise function and mechanism of HSP25 action has remained something of a mystery. That HSP25 falls into the class of proteins characterized as stress proteins is supported on many fronts. In particular, HSP25 has a similar distribution as some other stress proteins (HSP70 and HSP84).[98] Like these other proteins, HSP25 expression is induced by heat stress. HSP25 expression is also induced by other stressors,[99,100] including cold stress.[101] However, expression of HSP25 in response to stress is not a foregone result. A murine leukemic cell line appears to be incapable of expressing HSP25 despite apparent normal function of the promoter and the HSF1 transcription factor.[102] HSP25 expression is developmentally regulated, as well, being a predominant form of heat shock protein in many tissues.[103] The developmental regulation of expression appears to be of significant importance. Although expression levels are quite high in mature tissue, the expression level actually decreases during development of the rat heart, for example.[104] It appears that one function of HSP25 is as a chaperone protein similar to HSP70, trapping unfolded proteins in a folding competent state. As discussed previously, HSP25, HSP70, and HSP110 can directly interact with each other to form large complexes.[32] Using citrate synthase as a test protein, several reports propose a mechanism where HSP25 "catches" the denatured protein before it can irreversibly aggregate, allowing HSP70 to then initiate refolding of the protein.[94,105] The same mechanism appears to be true for denatured α-glucosidase, with αB-crystallin able to act similarly to HSP25.[94]

The ability of HSP25 to diminish irreversible aggregation of denatured proteins is interesting because HSP25 itself can exist in quite large oligomers. Ehrnsperger et al.[106] report that HSP25 exists in a hexadecameric structure that is in equilibrium with dimers and tetramers of the protein. In response to heat *in vivo*, large aggregates (heat granules) of the protein form whose functional ability appears to be retained.[106] Here is something of a dilemma.

As is discussed further below, HSP25 is a target of the mitogen-activated protein kinase-activated protein kinase 2 (MAPKAPK2). The activity of this kinase increases with stress. However, phosphorylation of HSP25 appears to diminish oligomerization and the protective effect of HSP25.[107] Therefore, it is not clear how oligomerization in response to heat stress coexists with activation of MAPKAPK2, unless it is to maintain an equilibrium between small and large oligomers. This is discussed further below.

HSP25 responds to a number of cell stressors. In fact, as with other classes of stress proteins, a cell stressor is identified by its ability to induce stress proteins.[12,108] For example, thermal stress causes an increase in the expression of HSP25 and the inducible form of HSP70 in the heart.[109] Although normally constitutively expressed, the hyperthermia-induced expression of HSP25 is anatomically selective. Whereas HSP70 is induced uniformly in the myocardium, HSP25 is induced in the left ventricle.[109] Translocation of HSP25 also takes place with thermal stress. In neonatal cardiomyocytes, heat stress causes translocation of HSP25 from the cytosol to sarcomeric structures,[110] and these have been identified as the actin I-bands.[109] In addition, HSP25 also translocates to the nucleus.[110] While these data indicate a cytoskeleton protective function for HSP25, overexpression of HSP25 was only protective in ischemic adult cardiomyocytes, not neonatal cells.[111] This is curious because thermal stress causes induction of HSP25 in neonatal cardiomyocytes.[110] However, it may be that αB-crystallin plays a larger protective role in the neonatal cells. As with HSP25, αB-crystallin translocates to the sarcomere in response to heat stress,[110] and overexpression of αB-crystallin is protective against ischemia in the neonatal cells.[111]

2. *Phosphorylation of HSP25*

Phosphorylation of HSP25 is a common stress response, although as mentioned above, the function of the phosphorylation is not clear. The protein is a potential target of several kinases and phosphatases, including PKA and PKC,[112] MAPKAPK2,[113] and calcium-calmodulin phosphatase 2B.[114] Consistent with MAPKAPK2 targeting HSP25, HSP25 is phosphorylated in a p38^{MAPK}-dependent manner in response to oxidative stress.[115-117] Similarly, heat shock,[115,118] osmotic stress,[119] shear stress,[120] and TNFα[118] all increase HSP25 phosphorylation through a p38^{MAPK}-MAPKAPK2 pathway. Whereas a p38^{MAPK}-mediated processes for induction of HSP70 appears to be necessary for the protective effect against osmotic stress,[121] it appears that protection is not a consequence of HSP25 phosphorylation. Several reports indicate that expression of mutant forms of the protein that cannot be phosphorylated still confer protection against heat and TNFα.[117,122] Similarly, blockade of p38^{MAPK} activity does not alter the protective effect of HSP25 against TNFα.[117] The phosphorylation may affect the dynamics of HSP25 aggregation and potential nuclear function, however. In response to TNFα, there is more pronounced aggregation of HSP25 in cells expressing non-phosphorylatable HSP25 mutants and in cells in which p38^{MAPK} is inhibited. These data indicate that phosphorylation may influence assembly and disassembly of the aggregates.

Another possibility is that the p38^{MAPK}/MAPKAPK2 stress response that targets HSP25 provides a transcription regulatory signal. Indeed, p38^{MAPK}/MAPKAPK2 also targets the serum response factor (SRF) transcriptional regulator.[123] Because HSP25 contains a nuclear targeting domain,[124] it is possible that phosphorylation regulates a nuclear role for the protein. However, although inhibition of the p38^{MAPK} pathway prevents HSP25 phosphorylation and differentiation of embryonic carcinoma cells into cardiomyocytes, it appears that the expression of HSP25, and not its phosphorylation, is important for the differentiation process.[125] Thus, the specific role of HSP25 phosphorylation remains to be determined.

B. HSP25, exercise, and protein synthesis

The association of HSP25 with the cytoskeleton raises a possibility that HSP25 may have other roles in addition to control of cytoskeletal structure. Indeed, it has been reported that HSP25 binds the adaptor protein eIF4γ and prevents assembly of the initiation complex.[1,126] This ability to diminish protein synthesis and HSP25's association with the cytoskeleton are consistent with the cytoskeleton's role in protein synthesis and the stress response.[127–134] For example, volume loading of the heart in the absence of a pressure load can be induced in rats by placing the animals in a head-down tilt. In these hearts, protein synthesis is rapidly depressed,[135] in part due to decreased initiation of protein synthesis.[10,56,136] In addition, there is a rapid erosion of the cytoskeleton in the cardiac myocytes.[137] We have found that HSP25 is rapidly induced in the volume-loaded rat heart in the absence of a pressure load. As shown in Figure 4.3, total expression of HSP25 protein increases within 8 hours of the volume loading. However, the expression level returns to unloaded levels by 18 hours and remains at control levels for at least 7 days of the head-down tilt. At 7 days, protein synthesis in the volume-loaded hearts remains depressed.[135] Association of HSP25 with the polysomes isolated from volume-loaded hearts also increases, lagging the total expression (Figure 4.3).

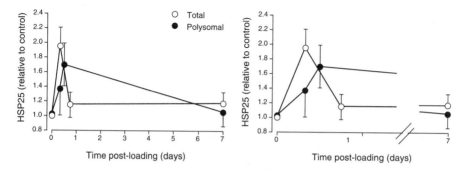

Figure 4.3   HSP25 is induced in rat hearts within 8 hours of volume loading (left; expanded time scale at right). Imposing a head-down tilt with tail-traction produces volume loading. The association of HSP25 protein with the polysomes isolated from the volume-loaded hearts reflects the induction of HSP25, lagging by several hours.

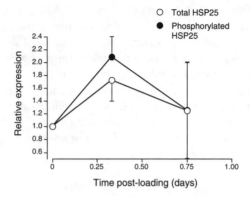

Figure 4.4   Along with induction of HSP25 protein expression by volume loading, HSP25 is also phosphorylated. On a relative basis, HSP25 is phosphorylated to a greater extent than the increase in protein expression.

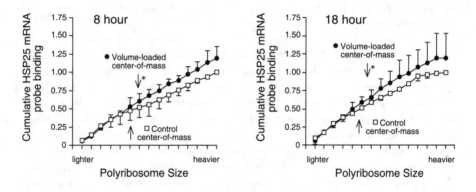

Figure 4.5   Consistent with the increased HSP25 protein expression, more HSP25 mRNA is associated with polysomes isolated from the volume-loaded hearts at 8 hours (left) and 18 hours (right). In contrast to the overall decreased protein synthesis rate and polysome size observed in the volume-loaded hearts,[10,135] there is a significant increase in the size of the HSP25 mRNA-containing polysomes (*$P < 0.05$, compared with control hearts).

This is consistent with total HSP25 expression driving the association with the polysomes. This association also returns to control levels with extended head-down tilt despite the maintenance of depressed protein synthesis. Phosphorylation of HSP25 mirrors the increase in HSP25 expression, although phosphorylation levels exceed the increased total HSP25 expression early in the volume loading (Figure 4.4).

Interestingly, the increased expression of HSP25 and its association with the polysomes at both 8 hours and 18 hours of volume loading is accompanied by an increase in the "center-of-mass" of the polysomes synthesizing HSP25 (Figure 4.5); center-of-mass is one measure of the dynamics of protein

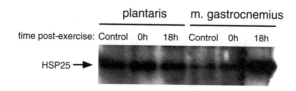

Figure 4.6   Plantaris muscle HSP25 protein expression increases immediately following running exercise (30 min, 20 m/min, 30% grade). In contrast, expression by the medial gastrocnemius muscle is delayed.

synthesis.[9,46] We have previously shown that the polysomes from the volume-loaded hearts indicate diminished initiation of protein synthesis with a shift of the center-of-mass to smaller polysomes.[10] This is attributed, in part, to an HSP70-driven phosphorylation of eIF2α.[56] However, despite the overall decrease in protein synthesis, but consistent with the increased expression of HSP25 protein, there is an increase in the HSP25 mRNA in polysomes isolated from the volume-loaded hearts (Figure 4.5). Furthermore, the HSP25 mRNA containing polysomes apparently escape the decreased initiation rate, as evidenced by their shift toward larger polysome sizes. Like volume-loading, which may impose an oxidative stress on the heart, running exercise also rapidly induces HSP25 expression. Thirty minutes of treadmill running by rats at 20 m/min produces an immediate post-exercise increase in HSP25 expression (Figure 4.6). Like the heart, and HSP70 levels in the plantaris (Figure 4.2), the plantaris HSP25 expression peaks early in the stress response and is easily detectable immediately following the exercise. Somewhat surprisingly, the medial gastrocnemius exhibits a later peak in HSP25 expression (Figure 4.6). This is in contrast to HSP70 expression in the medial gastrocnemius, which peaks immediately following the running exercise and returns toward control levels 18 hours post-exercise (Figure 4.2).

To summarize, both constitutively and inducibly expressed HSP25 may influence protein expression in muscle through several mechanisms. These include classical chaperone function, co-chaperone interaction and modulation of activity, and association with the cytoskeleton. However, the mechanisms of HSP25 function have considerable detail yet to be described.

IV. Conclusion

The discussion presented outlines some of the known mechanisms by which HSP70 and HSP25 function, and their potential impact upon protein synthesis in working muscle. Clearly, a major role for these stress proteins is as chaperones for unfolded or denatured proteins. In that role, the stress proteins may have direct or indirect effects on protein synthesis. However, the precise mechanisms are not always straightforward. This is perhaps not surprising, considering the multiple functions that have been identified for the stress proteins, both under stressed and nonstressed conditions. Much work remains to be done. Modulation of the expression of the stress proteins themselves is just one aspect of research on these proteins. Major consideration must also be given to modulation of function and activity of the stress proteins, especially under nonstressed conditions (as function may be different under these circumstances).

Acknowledgments

This work and manuscript preparation was supported by grants from the American Diabetes Association (DBT) and American Heart Association (DBT). The authors are indebted to Ms. Laura Malinick for her expert preparation of the figures.

References

1. Thomason, D. B., Translational control of gene expression in muscle, *Exerc. Sport Sci. Rev.*, 26, 165, 1998.
2. Hershey, J. W., Translational control in mammalian cells, *Annu. Rev. Biochem.*, 60, 717, 1991.
3. Moldave, K., Eukaryotic protein synthesis, *Annu. Rev. Biochem.*, 54, 1109, 1985.
4. Redpath, N. T. and Proud, C. G., Molecular mechanisms in the control of translation by hormones and growth factors, *Biochim. Biophys. Acta*, 1220, 147, 1994.
5. Ryazanov, A. G., Rudkin, B. B., and Spirin, A. S., Regulation of protein synthesis at the elongation stage. New insights into the control of gene expression in eukaryotes, *FEBS Lett.*, 285, 170, 1991.
6. Nakamura, Y., Ito, K., and Isaksson, L. A., Emerging understanding of translation termination, *Cell*, 82, 147, 1996.
7. Björnsson, A., Mottagui-Tabar, S., and Isaksson, L. A., Structure of the C-terminal end of the nascent peptide influences translation termination, *EMBO J.*, 15, 1696, 1996.
8. Locke, M., Noble, E. G., and Atkinson, B. G., Inducible isoform of HSP70 is constitutively expressed in a muscle fiber type specific pattern, *Am. J. Physiol.*, 261(Pt. 1), C774, 1991.
9. Ku, Z. et al., Decreased polysomal HSP70 may slow nascent polypeptide elongation during skeletal muscle atrophy, *Am. J. Physiol. (Cell Physiol.)*, 268, C1369, 1995.
10. Menon, V. et al., Decrease in heart peptide initiation during head-down tilt may be modulated by HSP70, *Am. J. Physiol. (Cell Physiol.)*, 268, C1375, 1995.
11. Locke, M. et al., Shifts in type I fiber proportion in rat hindlimb muscle are accompanied by changes in HSP72 content, *Am. J. Physiol.*, 266(Pt. 1), C1240, 1994.
12. Welch, W. J., Mammalian stress response: cell physiology, structure/function of stress proteins, and implications for medicine and disease, *Physiol. Rev.*, 72, 1063, 1992.
13. Craig, E. A., Chaperones: helpers along the pathways to protein folding, *Science*, 260, 1902, 1993.

14. Beckmann, R. P., Mizzen, L. E., and Welch, W. J., Interaction of HSP70 with newly synthesized proteins: implications for protein folding and assembly, *Science*, 248, 850, 1990.

15. Fink, A. L., Chaperone-mediated protein folding, *Physiol. Rev.*, 79, 425, 1999.

16. Brown, C. R. et al., The constitutive and stress inducible forms of HSP70 exhibit functional similarities and interact with one another in an ATP-dependent fashion, *J. Cell Biol.*, 120, 1101, 1993.

17. Brar, B. K. et al., Heat shock proteins delivered with a virus vector can protect cardiac cells against apoptosis as well as against thermal or hypoxic stress, *J. Mol. Cell. Cardiol.*, 31, 135, 1999.

18. Lau, S. S., Griffin, T. M., and Mestril, R., Protection against endotoxemia by HSP70 in rodent cardiomyocytes, *Am. J. Physiol. Heart Circ. Physiol.*, 278, H1439, 2000.

19. Liu, R. Y. et al., Expression of human HSP70 in rat fibroblasts enhances cell survival and facilitates recovery from translational and transcriptional inhibition following heat shock, *Cancer Res.*, 52, 3667, 1992.

20. Palleros, D. R., Welch, W. J., and Fink, A. L., Interaction of HSP70 with unfolded proteins: effects of temperature and nucleotides on the kinetics of binding, *Proc. Natl. Acad. Sci. U.S.A.*, 88, 5719, 1991.

21. Palleros, D. R. et al., ATP-induced protein-HSP70 complex dissociation requires K^+ but not ATP hydrolysis, *Nature*, 365, 664, 1993.

22. Palleros, D. R. et al., HSP70-protein complexes. Complex stability and conformation of bound substrate protein, *J. Biol. Chem.*, 269, 13107, 1994.

23. Greene, L. E. et al., Effect of nucleotide on the binding of peptides to 70-kDa heat shock protein, *J. Biol. Chem.*, 270, 2967, 1995.

24. McCarty, J. S. and Walker, G. C., DnaK as a thermometer: threonine-199 is site of autophosphorylation and is critical for ATPase activity, *Proc. Natl. Acad. Sci. U.S.A.*, 88, 9513, 1991.

25. Buchberger, A. et al., The chaperone function of DnaK requires the coupling of ATPase activity with substrate binding through residue E171, *EMBO J.*, 13, 1687, 1994.

26. McCarty, J. S. et al., The role of ATP in the functional cycle of the DnaK chaperone system, *J. Mol. Biol.*, 249, 126, 1995.

27. Szabo, A. et al., The ATP hydrolysis-dependent reaction cycle of the *Escherichia coli* HSP70 system DnaK, DnaJ, and GrpE, *Proc. Natl. Acad. Sci. U.S.A.*, 91, 10345, 1994.

28. Easton, D. P., Kaneko, Y., and Subjeck, J. R., The HSP110 and GRP1 70 stress proteins: newly recognized relatives of the HSP70s, *Cell Stress Chaperones*, 5, 276, 2000.

29. Yamagishi, N. et al., Modulation of the chaperone activities of HSC70/HSP40 by HSP105alpha and HSP105beta, *Biochem. Biophys. Res. Commun.*, 272, 850, 2000.

30. Raynes, D. A. and Guerriero, V., Isolation and characterization of isoforms of HSPBP1, inhibitors of HSP70, *Biochim. Biophys. Acta*, 1490, 203, 2000.

31. Raynes, D. A. and Guerriero, V., Inhibition of HSP70 ATPase activity and protein renaturation by a novel HSP70-binding protein, *J. Biol. Chem.*, 273, 32883, 1998.

32. Wang, X. Y. et al., Characterization of native interaction of HSP110 with HSP25 and HSC70, *FEBS Lett.*, 465, 98, 2000.

33. Minami, Y. and Minami, M., HSC70/HSP40 chaperone system mediates the HSP90-dependent refolding of firefly luciferase, *Genes Cells*, 4, 721, 1999.

34. Minami, Y. et al., A critical role for the proteasome activator PA28 in the HSP90-dependent protein refolding, *J. Biol. Chem.*, 275, 9055, 2000.

35. Nelson, R. J. et al., The translation machinery and 70 kd heat shock protein cooperate in protein synthesis, *Cell*, 71, 97, 1992.

36. Tai, P.-C. and Davis, B. D., Action of antibiotics on chain-initiating and on chain-elongating ribosomes., *Meth. Enz.*, 59, 851, 1979.

37. Eggers, D. K., Welch, W. J., and Hansen, W. J., Complexes between nascent polypeptides and their molecular chaperones in the cytosol of mammalian cells, *Mol. Biol. Cell*, 8, 1559, 1997.

38. Frydman, J. and Hartl, F. U., Principles of chaperone-assisted protein folding: differences between *in vitro* and *in vivo* mechanisms, *Science*, 272, 1497, 1996.

39. Frydman, J. et al., Folding of nascent polypeptide chains in a high molecular mass assembly with molecular chaperones, *Nature*, 370, 111, 1994.

40. Kudlicki, W. et al., Chaperone-dependent folding and activation of ribosome-bound nascent rhodanese. Analysis by fluorescence, *J. Mol. Biol.*, 244, 319, 1994.

41. Kudlicki, W. et al., The importance of the N-terminal segment for DnaJ-mediated folding of rhodanese while bound to ribosomes as peptidyl-tRNA, *J. Biol. Chem.*, 270, 10650, 1995.

42. Kudlicki, W. et al., Binding of an N-terminal rhodanese peptide to DnaJ and to ribosomes, *J. Biol. Chem.*, 271, 31160, 1996.

43. Tokatlidis, K. et al., Nascent chains: folding and chaperone interaction during elongation on ribosomes, *Philos. Trans. R. Soc. Lond. B., Biol. Sci.*, 348, 89, 1995.

44. Craig, E. A. and Jacobsen, K., Mutations of the heat inducible 70 kilodalton genes of yeast confer temperature sensitive growth, *Cell*, 38, 841, 1984.

45. Brodsky, J. L., Lawrence, J. G., and Caplan, A. J., Mutations in the cytosolic DnaJ homologue, YDJ1, delay and compromise the efficient translation of heterologous proteins in yeast, *Biochemistry*, 37, 18045, 1998.

46. Ku, Z. and Thomason, D. B., Soleus muscle nascent polypeptide chain elongation slows protein synthesis rate during non-weightbearing, *Am. J. Physiol. (Cell Physiol.)*, 267, C115, 1994.

47. Gambill, B. D. et al., A dual role for mitochondrial heat shock protein 70 in membrane translocation of preproteins, *J. Cell Biol.*, 123, 109, 1993.

48. Kang, P. J. et al., Requirement for HSP70 in the mitochondrial matrix for translocation and folding of precursor proteins, *Nature*, 348, 137, 1990.

49. Voos, W. et al., Mechanisms of protein translocation into mitochondria, *Biochim. Biophys. Acta*, 1422, 235, 1999.

50. Brodsky, J. L. et al., Mitochondrial HSP70 cannot replace BiP in driving protein translocation into the yeast endoplasmic reticulum, *FEBS Lett.*, 435, 183, 1998.

51. Brodsky, J. L. et al., The requirement for molecular chaperones during endoplasmic reticulum-associated protein degradation demonstrates that protein export and import are mechanistically distinct, *J. Biol. Chem.*, 274, 3453, 1999.

52. Hershey, J. W., Overview: phosphorylation and translation control, *Enzyme*, 44, 17, 1990.

53. Matts, R. L. and Hurst, R., The relationship between protein synthesis and heat shock proteins levels in rabbit reticulocyte lysates, *J. Biol. Chem.*, 267, 18168, 1992.

54. Matts, R. L. et al., Interactions of the heme-regulated eIF-2 alpha kinase with heat shock proteins in rabbit reticulocyte lysates, *J. Biol. Chem.*, 267, 18160, 1992.

55. Matts, R. L., Hurst, R., and Xu, Z., Denatured proteins inhibit translation in hemin-supplemented rabbit reticulocyte lysate by inducing the activation of the heme-regulated eIF-2 alpha kinase, *Biochemistry,* 32, 7323, 1993.

56. Menon, V. and Thomason, D. B., Head-down tilt increases rat cardiac muscle eIF-2α phosphorylation, *Am. J. Physiol. (Cell Physiol.),* 269, C802, 1995.

57. Someren, J. S. et al., Heat shock proteins 70 and 90 increase calcineurin activity *in vitro* through calmodulin-dependent and independent mechanisms, *Biochem. Biophys. Res. Commun.,* 260, 619, 1999.

58. Henics, T. et al., Mammalian HSP70 and HSP110 proteins bind to RNA motifs involved in mRNA stability, *J. Biol. Chem.,* 274, 17318, 1999.

59. Scandurro, A. B. et al., Interaction of erythropoietin RNA binding protein with erythropoietin RNA requires an association with heat shock protein 70, *Kidney Int.,* 51, 579, 1997.

60. Okada, S. et al., HSP70 and ribosomal protein L2: novel 5S rRNA binding proteins in *Escherichia coli, FEBS Lett.,* 485, 153, 2000.

61. Blake, M. J. et al., Blood pressure and heat shock protein expression in response to acute and chronic stress, *Hypertension,* 25(4 Pt. 1), 539, 1995.

62. Hamet, P. et al., Heat stress genes in hypertension, *J. Hypertens. Suppl.,* 8, S47, 1990.

63. Patton, W. F. et al., Components of the protein synthesis and folding machinery are induced in vascular smooth muscle cells by hypertrophic and hyperplastic agents. Identification by comparative protein phenotyping and microsequencing, *J. Biol. Chem.,* 270, 21404, 1995.

64. Xu, Q. et al., Mechanical stress-induced heat shock protein 70 expression in vascular smooth muscle cells is regulated by Rac and Ras small G proteins but not mitogen-activated protein kinases, *Circ. Res.,* 86, 1122, 2000.

65. Bornfeldt, K. E., Stressing Rac, Ras, and downstream heat shock protein 70, *Circ. Res.,* 86, 1101, 2000.

66. Sheikh-Hamad, D. et al., p38 kinase activity is essential for osmotic induction of mRNAs for HSP70 and transporter for organic solute betaine in Madin-Darby canine kidney cells, *J. Biol. Chem.,* 273, 1832, 1998.

67. Bates, P. C. et al., Exercise and muscle protein turnover in the rat, *J. Physiol. Lond.,* 303, 41P, 1980.

68. Dohm, G. L. et al., Effect of exercise on synthesis and degradation of muscle protein, *Biochem. J.,* 188, 255, 1980.

69. Dohm, G. L. et al., Measurement of *in vivo* protein synthesis in rats during an exercise bout, *Biochem. Med.,* 27, 367, 1982.

70. Farrell, P. A. et al., Regulation of protein synthesis after acute resistance exercise in diabetic rats, *Am. J. Physiol.,* 276(4 Pt. 1), E721, 1999.

71. Gautsch, T. A. et al., Availability of eIF4E regulates skeletal muscle protein synthesis during recovery from exercise, *Am. J. Physiol.,* 274(2 Pt. 1), C406, 1998.

72. Farrell, P. A. et al., Eukaryotic initiation factors and protein synthesis after resistance exercise in rats, *J. Appl. Physiol.,* 88, 1036, 2000.

73. Hernandez, J. M., Fedele, M. J., and Farrell, P. A., Time course evaluation of protein synthesis and glucose uptake after acute resistance exercise in rats, *J. Appl. Physiol.,* 88, 1142, 2000.

74. Bylund-Fellenius, A. C. et al., Protein synthesis versus energy state in contracting muscles of perfused rat hindlimb, *Am. J. Physiol.,* 246(4 Pt. 1), E297, 1984.

75. Hernando, R. and Manso, R., Muscle fibre stress in response to exercise: synthesis, accumulation and isoform transitions of 70-kDa heat-shock proteins, *Eur. J. Biochem.*, 243, 460, 1997.

76. Kilgore, J. L., Musch, T. I., and Ross, C. R., Physical activity, muscle, and the HSP70 response, *Can. J. Appl. Physiol.*, 23, 245, 1998.

77. Liu, Y. et al., Human skeletal muscle HSP70 response to training in highly trained rowers, *J. Appl. Physiol.*, 86, 101, 1999.

78. Liu, Y. et al., Human skeletal muscle HSP70 response to physical training depends on exercise intensity, *Int. J. Sports. Med.*, 21, 351, 2000.

79. Locke, M., Noble, E. G., and Atkinson, B. G., Exercising mammals synthesize stress proteins, *Am. J. Physiol.*, 258(4 Pt. 1), C723, 1990.

80. Locke, M. and Noble, E. G., Stress proteins: the exercise response, *Can. J. Appl. Physiol.*, 20, 155, 1995.

81. Puntschart, A. et al., HSP70 expression in human skeletal muscle after exercise, *Acta Physiol. Scand.*, 157, 411, 1996.

82. Locke, M. et al., Activation of heat-shock transcription factor in rat heart after heat shock and exercise, *Am. J. Physiol.*, 268(6 Pt. 1), C1387, 1995.

83. Sjodin, B., Hellsten Westing, Y., and Apple, F. S., Biochemical mechanisms for oxygen free radical formation during exercise, *Sports Med.*, 10, 236, 1990.

84. Sahlin, K., Ekberg, K., and Cizinsky, S., Changes in plasma hypoxanthine and free radical markers during exercise in man, *Acta Physiol. Scand.*, 142, 275, 1991.

85. Donovan, C. M. and Faulkner, J. A., Plasticity of skeletal muscle: regenerating fibers adapt more rapidly than surviving fibers, *J. Appl. Physiol.*, 62, 2507, 1987.

86. Armstrong, R. B., Initial events in exercise-induced muscular injury, *Med. Sci. Sports Exerc.*, 22, 429, 1990.

87. Samelman, T. R., Heat shock protein expression is increased in cardiac and skeletal muscles of Fischer 344 rats after endurance training, *Exp. Physiol.*, 85, 92, 2000.

88. Demirel, H. A. et al., The effects of exercise duration on adrenal HSP72/73 induction in rats, *Acta Physiol. Scand.*, 167, 227, 1999.

89. Fehrenbach, E. et al., HSP expression in human leukocytes is modulated by endurance exercise, *Med. Sci. Sports Exerc.*, 32, 592, 2000.

90. Fehrenbach, E. et al., Transcriptional and translational regulation of heat shock proteins in leukocytes of endurance runners, *J. Appl. Physiol.*, 89, 704, 2000.

91. Gonzalez, B., Hernando, R., and Manso, R., Stress proteins of 70 kDa in chronically exercised skeletal muscle, *Pflügers Arch.*, 440, 42, 2000.

92. http://ash.gene.ucl.ac.uk/nomenclature/.

93. http://bioinformatics.weizmann.ac.il/cards/.

94. Jakob, U. et al., Small heat shock proteins are molecular chaperones, *J. Biol. Chem.*, 268, 1517, 1993.

95. Merck, K. B. et al., Structural and functional similarities of bovine alpha-crystallin and mouse small heat-shock protein. A family of chaperones, *J. Biol. Chem.*, 268, 1046, 1993.

96. Atomi, Y., Yamada, S., and Nishida, T., Early changes of alpha B-crystallin mRNA in rat skeletal muscle to mechanical tension and denervation, *Biochem. Biophys. Res. Commun.*, 181, 1323, 1991.

97. Atomi, Y. et al., Fiber-type-specific alphaB-crystallin distribution and its shifts with T(3) and PTU treatments in rat hindlimb muscles, *J. Appl. Physiol.*, 88, 1355, 2000.

98. Tanguay, R. M., Wu, Y., and Khandjian, E. W., Tissue-specific expression of heat shock proteins of the mouse in the absence of stress, *Dev. Genet.*, 14, 112, 1993.

99. Mattei, E. et al., Induction of stress proteins in murine and human melanoma cell cultures, *Tumori*, 72, 129, 1986.

100. Kim, Y. J. et al., Arsenate induces stress proteins in cultured rat myoblasts, *J. Cell Biol.*, 96, 393, 1983.

101. Laios, E., Rebeyka, I. M., and Prody, C. A., Characterization of cold-induced heat shock protein expression in neonatal rat cardiomyocytes, *Mol. Cell. Biochem.*, 173, 153, 1997.

102. Neininger, A. and Gaestel, M., Evidence for a HSP25-specific mechanism involved in transcriptional activation by heat shock, *Exp. Cell. Res.*, 242, 285, 1998.

103. Gernold, M. et al., Development and tissue-specific distribution of mouse small heat shock protein HSP25, *Dev. Genet.*, 14, 103, 1993.

104. Lutsch, G. et al., Abundance and location of the small heat shock proteins HSP25 and αB-crystallin in rat and human heart, *Circulation*, 96, 3466, 1997.

105. Ehrnsperger, M. et al., Binding of non-native protein to HSP25 during heat shock creates a reservoir of folding intermediates for reactivation, *EMBO J.*, 16, 221, 1997.

106. Ehrnsperger, M. et al., The dynamics of HSP25 quaternary structure. Structure and function of different oligomeric species, *J. Biol. Chem.*, 274, 14867, 1999.

107. Rogalla, T. et al., Regulation of HSP27 oligomerization, chaperone function, and protective activity against oxidative stress/tumor necrosis factor alpha by phosphorylation, *J. Biol. Chem.*, 274, 18947, 1999.

108. Morimoto, R. I., Cells in stress: transcriptional activation of heat shock genes, *Science*, 259, 1409, 1993.

109. Hoch, B. et al., HSP25 in isolated perfused rat hearts: localization and response to hyperthermia, *Mol. Cell. Biochem.*, 160, 231, 1996.

110. van de Klundert, F. A. et al., alpha B-crystallin and HSP25 in neonatal cardiac cells — differences in cellular localization under stress conditions, *Eur. J. Cell. Biol.*, 75, 38, 1998.

111. Martin, J. L. et al., Small heat shock proteins and protection against ischemic injury in cardiac myocytes, *Circulation*, 96, 4343, 1997.

112. Gaestel, M. et al., Identification of the phosphorylation sites of the murine small heat shock protein HSP25, *J. Biol. Chem.*, 266, 14721, 1991.

113. Stokoe, D. et al., Identification of MAPKAP kinase 2 as a major enzyme responsible for the phosphorylation of the small mammalian heat shock proteins, *FEBS Lett.*, 313, 307, 1992.

114. Gaestel, M. et al., Dephosphorylation of the small heat shock protein HSP25 by calcium/calmodulin-dependent (type 2B) protein phosphatase, *J. Biol. Chem.*, 267, 21607, 1992.

115. Zu, Y. L. et al., High expression and activation of MAP kinase-activated protein kinase 2 in cardiac muscle cells, *J. Mol. Cell. Cardiol.*, 29, 2159, 1997.

116. Clerk, A., Michael, A., and Sugden, P. H., Stimulation of multiple mitogen-activated protein kinase sub-families by oxidative stress and phosphorylation of the small heat shock protein, HSP25/27, in neonatal ventricular myocytes, *Biochem. J.*, 333(Pt. 3), 581, 1998.

117. Preville, X. et al., Analysis of the role of HSP25 phosphorylation reveals the importance of the oligomerization state of this small heat shock protein in its protective function against TNFalpha- and hydrogen peroxide-induced cell death, *J. Cell. Biochem.*, 69, 436, 1998.

118. Engel, K. et al., MAPKAP kinase 2 is activated by heat shock and TNF-alpha: *in vivo* phosphorylation of small heat shock protein results from stimulation of the MAP kinase cascade, *J. Cell. Biochem.*, 57, 321, 1995.

119. Muller, E. et al., Effects of long-term changes in medullary osmolality on heat shock proteins HSP25, HSP60, HSP72 and HSP73 in the rat kidney, *Pflügers Arch.*, 435, 705, 1998.

120. Azuma, N. et al., Role of p38 MAP kinase in endothelial cell alignment induced by fluid shear stress, *Am. J. Physiol. Heart Circ. Physiol.*, 280, H189, 2001.

121. Beck, F. X. et al., Heat shock proteins and the cellular response to osmotic stress, *Cell. Physiol. Biochem.*, 10, 303, 2000.

122. Knauf, U. et al., Stress- and mitogen-induced phosphorylation of the small heat shock protein HSP25 by MAPKAP kinase 2 is not essential for chaperone properties and cellular thermoresistance, *EMBO J.*, 13, 54, 1994.

123. Heidenreich, O. et al., MAPKAP kinase 2 phosphorylates serum response factor *in vitro* and *in vivo*, *J. Biol. Chem.*, 274, 14434, 1999.

124. Engel, K., Plath, K., and Gaestel, M., The MAP kinase-activated protein kinase 2 contains a proline-rich SH3-binding domain, *FEBS Lett.*, 336, 143, 1993.

125. Davidson, S. M. and Morange, M., HSP25 and the p38 MAPK pathway are involved in differentiation of cardiomyocytes, *Dev. Biol.*, 218, 146, 2000.

126. Cuesta, R., Laroia, G., and Schneider, R. J., Chaperone HSP27 inhibits translation during heat shock by binding eIF4G and facilitating dissociation of cap-initiation complexes, *Genes Dev.*, 14, 1460, 2000.

127. Bektas, M. et al., Interactions of eukaryotic elongation factor 2 with actin: a possible link between protein synthetic machinery and cytoskeleton, *FEBS Lett.*, 356, 89, 1994.

128. Durso, N. A. and Cyr, R. J., Beyond translation: elongation factor-1 alpha and the cytoskeleton, *Protoplasma*, 180, 99, 1994.

129. Hesketh, J. E. and Pryme, I. F., Interaction between mRNA, ribosomes and the cytoskeleton, *Biochem. J.*, 277(Pt. 1), 1, 1991.

130. Hesketh, J., Translation and the cytoskeleton: a mechanism for targeted protein synthesis, *Mol. Biol. Rep.*, 19, 233, 1994.

131. van Bergen en Henegouwen, P. M. et al., Studies on a possible relationship between alterations in the cytoskeleton and induction of heat shock protein synthesis in mammalian cells, *Int. J. Hyperthermia*, 1, 69, 1985.

132. Zhu, Y. et al., Phosphorylated HSP27 associates with the activation-dependent cytoskeleton in human platelets, *Blood*, 84, 3715, 1994.

133. Hovland, R., Hesketh, J. E., and Pryme, I. F., The compartmentalization of protein synthesis: importance of cytoskeleton and role in mRNA targeting, *Int. J. Biochem. Cell. Biol.*, 28, 1089, 1996.

134. Condeelis, J., Elongation factor 1 alpha, translation and the cytoskeleton, *Trends Biochem. Sci.*, 20, 169, 1995.

135. Thomason, D. B., Biggs, R. B., and Booth, F. W., Protein metabolism and β-myosin heavy-chain mRNA in unweighted soleus muscle, *Am. J. Physiol.*, 257, R300, 1989.

136. Takala, T., Protein synthesis in the isolated perfused rat heart, *Basic Res. Cardiol.*, 76, 44, 1981.

137. Thomason, D. B., Anderson III, O., and Menon, V., Fractal analysis of cytoskeleton rearrangement in rat cardiac muscle during head-down tilt, *J. Appl. Physiol.*, 81, 1522, 1996.

chapter five

Stress proteins and myocardial protection

Joseph W. Starnes

Contents

I. Introduction

It is now well accepted that chronic physical activity lowers the risk of cardiovascular disease and reduces deaths from heart attacks.[1-4] The American Heart Association now categorizes lack of physical activity as a risk factor for cardiovascular disease, which is the leading cause of death in the United States. Therefore, appropriate exercise has tremendous potential for reducing health care costs in this country. Consequently, there is a great deal of interest in determining such key issues as the underlying mechanisms for improved cardiovascular health and the amount of exercise necessary to achieve it. One indication of the health of a heart is its ability to tolerate and recover from an acute severe stress. Many investigations have focused on the intrinsic ability of the heart from exercise trained animals to tolerate ischemia. In addition to being clinically relevant, ischemia is a powerful tool to probe basic myocardial physiology. Several studies, using various models and endpoint measures, have reported that exercise results in improved cardioprotection against imposed ischemia and subsequent reperfusion, that is, return of blood flow to the coronary circulation. Another intervention that leads to protection against reperfusion injury is a bout of heat stress. It is generally believed that the induction of certain cardioprotective proteins is responsible for the adaptation to both interventions and there appears to be many similarities between exercise-induced and heat shock-induced cardioprotection. Proteins that are strong candidates include, but are not limited to, a stress-induced member of the 70-kDalton heat shock protein family (HSP70) and the antioxidant enzymes superoxide dismutase (SOD) and catalase. This chapter focuses primarily on the role of these proteins because of the considerable amount of attention they have received. We discuss their possible roles in both heat shock-induced and exercise-induced protection against reperfusion injury.

II. Types of reperfusion injury

Before discussing cardioprotective proteins, a definition of reperfusion injury and explanation of experimental models used will be helpful. Reperfusion injury represents the reversible and irreversible injury caused by the return of blood flow (reperfusion) to a previously ischemic heart. While reperfusion is absolutely necessary to prevent cell death that would eventually occur during ischemia, the return of oxygen causes damage of its own. The specific type and amount of damage is related to the length and/or severity of the preceding ischemic period. Reperfusion following brief ischemic periods (20 minutes or less) usually produces temporary postischemic dysfunction that is fully reversible within a few minutes to a few hours. This dysfunction has been given the term "myocardial stunning" by Braunwald and Kloner[5] and is characterized by the absence of irreversible damage and the maintenance of normal coronary flow.[6] With more prolonged or severe ischemia, irreversible necrosis and cell death occur, leading to myocardial infarction. Necrosis is characterized by disruption of cell membranes and can be detected during reperfusion by observing the quantity of cellular enzymes and other proteins released into the coronary effluent. The actual amount of infarcted tissue (infarct size) can be determined by administering dyes that discriminate between viable and nonviable cells.

Stunning can also be induced during exercise. Exercise results in significant increases in myocardial oxygen demand and if oxygen supply is limited by a stenosis or coronary spasm, ischemia and dysfunction may occur. Dysfunction persists for some time after exercise, although oxygen demand no longer exceeds supply.[7,8] In hypertrophied hearts, exercise has been reported to produce postischemic stunning even without evidence of coronary stenosis.[9] The observations that myocardial stunning can also occur in exercise situations has major clinical implications.

III. Experimental models

A diverse array of models have been developed and used to study ischemia-reperfusion injury (for reviews, see References 10 and 11). Most models were devised to answer specific questions and thus often cannot be used to study other aspects of reperfusion injury. A close examination of the literature reveals that many of the apparently conflicting conclusions regarding reperfusion injury are likely due to differences in the model systems employed. Most studies investigating the effect of exercise on reperfusion injury have been carried out on one species, the rat, and have used either an *in vivo* or *in situ* model of regional ischemia or an isolated perfused heart preparation. In the *in vivo/in situ* models, ischemia to a portion of the left ventricle is produced by occluding a coronary artery for various lengths of time before allowing blood to return. Lack of an effective coronary collateral circulation in the rat heart allows for a well-defined ischemic region. The advantage of this model is that it allows for normal tissue and blood interactions, and the response of the heart to the ischemic event can be followed for several days after reperfusion. This model is excellent for evaluation of infarct size and ventricular remodeling because it is very stable and receives all the appropriate nutrients and growth factors from the host. Contractile measures of whole ventricle are not very sensitive indicators of myocardial injury because these measures can be altered by neural and hormonal input from the host, who may have a different short-term goal than the investigator. This input could result in changes in afterload, preload, and/or contractility compared to baseline or preischemia values. Also, because the heart is usually performing at a low workload in an anaesthetized animal and only a portion of the ventricle is actually injured, healthy tissue may be able to "take up some of the slack" for the dysfunctional area. However, regional contractile function can be estimated with the aid of piezoelectric crystals or other sophisticated instrumentation.

In isolated perfused preparations, the heart is removed from the animal and rapidly secured by the aorta to an apparatus containing a physiologically balanced perfusion solution. Once on the apparatus, a good preparation is stable for up to 3 hours. The entire heart can be made ischemic (global ischemia), providing homogeneous injury throughout the ventricles and negating the potential contribution of collateral blood supply. The perfusion solution does not normally contain red blood cells, but is gassed with a 95% O_2:5% CO_2 mixture to maintain oxygen content above 600 torr and pH at 7.4. Coronary flow can be conveniently collected and used to measure various metabolic products and markers of myocardial damage, including cytosolic enzyme/protein leakage. There are two variations of the isolated perfused model: the Langendorff preparation and the working heart preparation. The most commonly used model is the Langendorff preparation, originally described in 1895,[12] in which the heart is perfused retrogradely through the aorta at a constant flow or pressure. The perfusate enters the coronary ostia located just above the aortic valve and is pushed through the coronary circulation by the imposed perfusion pressure. Raising the perfusion pressure increases mechanical activity by the "garden hose" effect in which the arteries distend and stiffen the myocardial walls requiring more energy during contraction.[13,14] Typical mechanical measures include intraventricular systolic and diastolic pressures, maximum rates of pressure development and relaxation (±dP/dt), and the product of heart rate and developed pressure (rate–pressure product). Usually these mechanical measures are made with the aid of a balloon inserted into the left ventricle to assure constant ventricular volume. This preparation is also referred to as the nonworking heart because it does not perform external work (i.e., cardiac output). However, it does perform metabolic and mechanical work that is dependent on its own contractile properties and the perfusion pressure.

The working heart preparation was described in 1967 by Neely et al.[15] In this preparation, both the aorta and the left atrium are cannulated and the perfusate traverses the heart as it would *in vivo*. The ventricle pumps perfusate into the aorta and pushes the fluid up a column that is high enough to then provide the perfusion pressure necessary to drive the perfusate through the coronary tree. If the ventricle cannot push the aortic flow high enough up the column, perfusion pressure to the coronary tree falls and the heart fails. Compared to the Langendorff preparation, the working heart preparation provides greater sensitivity to detecting small changes in contractility. It is the model of choice for studying postischemic stunning.[11] However, it cannot be used for evaluating reperfusion injury following prolonged or severe ischemic events because the ventricle will suffer too much irreversible damage and be unable to pump with enough force to adequately perfuse the coronary tree. For such studies, the Langendorff preparation is used because the perfusion apparatus ensures adequate perfusion pressure.

IV. Heat shock and cardioprotection

Heat shock proteins received their names because they were originally discovered as a stress response to sublethal elevations of core temperature.[16,17] Typically, the numerous and diverse heat shock or heat stress proteins are differentiated and named by molecular weight. It is now known that stress proteins are induced by a wide variety of stressors and can provide protection against different subsequent stresses. Protection provided in this manner has been termed "cross-tolerance." Li and Hahn[18] were perhaps the first to conduct experiments relating to this concept. They reported that pretreatment with ethanol produces thermotolerance much like that observed with prior heat shock. Evidence that prior heat shock could protect the cell against a variety of stresses led Currie et al.[19] to determine if prior heat shock could provide protection against ischemia reperfusion injury in the myocardium. In this 1988 study, they used a heating pad to raise core temperatures of rats to 42°C for 15 minutes. Hearts were isolated 24 hours later and subjected to 30 minutes of low-flow (1 ml/min) ischemia followed by 30 minutes of reperfusion using the Langendorff method. Hearts from animals that were previously heat shocked had significantly greater recovery of force and rates of contraction and relaxation compared to controls, thus indicating that whole-body heat shock could result in intrinsic myocardial protection from ischemia.

A majority of the literature supports the findings of Currie et al.[19] that heat shock provides cardioprotection. Other investigations have found better recovery of function and less enzyme leakage following an ischemic bout.[20–24] In addition, a number of studies have also demonstrated a decrease in infarct size following prior treatment with heat shock.[25–28] A few investigations, however, have not observed improved cardioprotection following heat shock.[25,26,29] The reasons for these discrepancies appear to be the model used to assess cardioprotection as well as the severity of the insult being administered. For example, Donnelly et al.[25] observed reductions in infarct size following 35 minutes of regional ischemia but not following 45 minutes of regional ischemia. The authors indicated that 45 minutes of ischemia may have been too severe to be overcome by the protective effects of prior heat shock. These data suggest that prior heat stress will attenuate ischemic damage for a certain window of time after an initial threshold of damage.

A. HSP70

Although it appears clear that prior heat stress can provide protection against ischemic injury, the mechanisms involved are less distinct. In the original study by Currie et al.,[19] the authors suggested that certain protective proteins were expressed following heat shock, especially the one they called SP71. SP71, more commonly referred to as HSP72 or HSP70, is the major stress-inducible form of the HSP70 family of stress proteins. The HSP70 family of proteins is the most highly conserved of the stress proteins from bacteria to man. Most eukaryotes contain over a dozen different members of this family.[30] The high degree of conservation of these proteins implies a vital and universal function and a majority of the literature on heat shock-induced "tolerance" have focused on these proteins. HSP70/72 and its constitutive homolog HSC70 (or HSP73) fall under the classification of molecular chaperones. Molecular chaperones, and HSP70/72 in particular, are involved in the proper folding of newly synthesized proteins.[30,31] In addition, HSP70/72 has been shown to be involved in protein transport, disassembly and degradation, as well as preventing protein aggregation and denaturation. It may also help restore the function of proteins that are damaged during stress.[32] Other possible roles are be discussed later in this chapter. We use the term "HSP70" throughout this chapter to refer to the stress-inducible form of the HSP70 family.

Within 2 years after the discovery by Currie et al.,[19] the same group took a closer look at the potential relationship between HSP70 expression and cardioprotection.[21] This study followed recovery of force and amount of enzyme leakage after global ischemia in Langendorff perfused hearts at 24, 48, 96, and 192 hours post heat shock. The results indicated a similar rise and fall in cardioprotection and HSP70 expression. However, they also found an increase in catalase activity and, when it was inactivated, the hyperthermia-induced enhancement of postischemic recovery was abolished. Subsequently, Hutter et al.,[33] employing 35 minutes of coronary artery occlusion in the intact rat, demonstrated a direct correlation 24 hours after heat shock between the level of HSP70 and post-ischemic infarct size ($R^2 = 0.97$). Heating the animals at different temperatures was used to alter the induction of heat shock protein. In contrast, Qian et al.,[26] using a similar *in vivo* ischemia-reperfusion protocol as Hutter et al.,[33] reported a disassociation between HSP70 expression and infarct size following heat shock. They induced 30 minutes of regional ischemia at 2, 4, 12, 24, and 30 hours following whole-body heat shock. Only the 24- and 30-hour groups had significant reductions in infarct size after 90 minutes of reperfusion, but HSP70 expression reached 80% of its maximum value at 4 hours and a maximum value at 12 hours. Furthermore, no differences in mean arterial pressures or rate pressure product were observed between any group at any time. However, these are not very sensitive measures of cardiac function. Overall, it is likely that heat shock induces several proteins that may also contribute to the cardioprotective effect.

Although it has been difficult to establish that HSP70 is the primary stress protein involved in cardioprotection following heat shock, there is now little doubt that it is a potent cardioprotective agent. This has been established using transgenic animal models, which allow for the evaluation of the protective effect of HSP70 independent of most other factors. Hutter et al.[34] found reduced infarct size following 30 minutes of coronary artery ligation *in vivo* in mice overexpressing the inducible rat HSP70. Studies utilizing global ischemia in the isolated perfused Langendorff model have also found HSP70 to be protective. Marber et al.[35] showed that transgenic mice overexpressing the inducible rat HSP70 had a 40% reduction in infarct size, 50% reduction in creatine kinase release, and a twofold increase in contractile function compared to controls following 20 minutes of global ischemia and 30 minutes of reperfusion. Plumier et al.[36] and Radford et al.[37] also found improved cardioprotection in mice overexpressing the human inducible form of HSP70. Trost et al.[38] looked at functional recovery of transgenic mouse hearts following a brief period of ischemia using both *in vitro* and *in vivo* models. The *in vitro* model consisted of 10 minutes of global ischemia in isolated Langendorff perfused hearts, and the *in vivo* model consisted of 8 minutes of regional ischemia by coronary artery ligation. The ischemic periods were shown not to lead to necrosis, but the authors hesitate to classify them as stunning. In the *in vivo* experiments, a regional epicardial strain technique was used that provided a more sensitive measure of changes in regional myocardial function compared to measures of ventricular pressure development or rate pressure product. The HSP70-positive mice were found to have better recovery upon reperfusion in both models.

In transgenic models, the heart is genetically altered to overexpress HSP70 from birth. Therefore, it may adapt over time to develop other

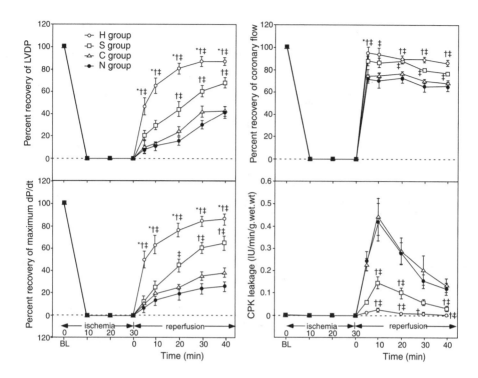

Figure 5.1   Evidence that heat stress and elevating HSP70 protect against ischemia-reperfusion injury. Isolated hearts were subjected to 30 minutes of global ischemia. Values are means ± SE. Upper left, left ventricular developed pressure (LVDP); lower left; maximum rate of pressure development (dP/dt); upper right, coronary flow; lower right; creatine phosphokinase (CPK) leakage. H group, transfected with human HSP70 gene; S group, heated to 42°C rectal temperature for 20 minutes; C group, sham control; N group, untreated control; BL, baseline pre-ischemic value. Overall, heat shocked group recovered better than controls and HSP70 transfected group generally recovered better than all groups. *P < 0.05 vs. S group; †P <0.05 vs. C group; ‡P < 0.05 vs. N group. (From Suzuki, K. et al., *J. Clin. Invest.*, 99, 1645, 1997. With permission.)

defenses not found in the natural heart. Another genetic manipulation is the process of gene transfection, where a gene is carried by a virus to the host's DNA. Suzuki et al.[39] transfected human HSP70 into rat hearts to produce a 12- to 20-fold increase in 4 days, which was double the amount induced by heat shock (20 minutes at 42°C, 24 hours prior to determination). As in other studies, protection from 30 minutes of global ischemia and subsequent reperfusion in the Langendorff perfused heart was improved by heat shock, but gene transfected hearts provided an even greater magnitude of protection. Key findings of this study are displayed in Figure 5.1. The overall results led the authors to conclude "that HSP70 is by far the most powerful and effective mechanism for myocardial protection among many endogenous protective mechanisms mobilized by heat stress." Gene transfected rat hearts have also been found to suffer less apoptosis compared to non-transfected controls using both *in vitro* and *in vivo* ischemia-reperfusion models.[40] Overall, these studies with transgenic and transfected animals clearly demonstrate that HSP70 is one factor in myocardial protection from ischemia-reperfusion injury.

The mechanisms by which HSP70 is cardioprotective have not been fully elucidated. It is known to act as a molecular chaperone, which may facilitate the restoration of enzymes and other key proteins denatured during ischemic stress.[30–32] Also, Kawana et al.[41] have recently provided evidence that HSP70 can protect the heart by attenuating the activity of nuclear poly(ADP-ribose) synthetase, which consumes excessive amounts of energy during reoxygenation, thereby making an insufficient amount available for pump function. Consistent with this, Knowlton[42] has recently carried out an experiment demonstrating that nuclear accumulation of HSP70 is required to confer protection. Using transfected rats from the same colony that was used in Figure 5.1, Sakaguchi et al.[43] demonstrated that the cardioprotective effects of HSP70 are mediated by ecto-5'-nucleotidase, which catalyzes the breakdown of AMP to adenosine. This enzyme is significantly increased by overexpression of HSP70, resulting in a corresponding increase in adenosine release (approximately fivefold) during reperfusion. Inhibiting the enzyme in transfected hearts abolished the HSP70-induced improvement in postischemic recovery and decrease in enzyme leakage, but had little effect in control hearts.

Several groups have explored protein kinase C (PKC) along with HSP70 in the cardioprotection induced by heat stress.[44–46] These studies used a variety of models, ranging from isolated cells to regional ischemia *in vivo*. All studies found that PKC inhibitors administered prior to heat stress did not prevent HSP70 induction but abolished the cardioprotective response normally observed within 24 to 48 hours. Thus, heat-stress induced cardioprotection was dissociated from HSP70 induction. The possibility exists that PKC inhibitors affect HSP70 function, perhaps by preventing post-translational phosphorylation of the protein.[44] The relationship between PKC and HSP70 synthesis and function is still unresolved. By the time this chapter is published, there will likely be several more papers in print exploring the cardioprotective mechanism of HSP70.

B.   Catalase and superoxide dismutase

It is now widely accepted that oxygen-derived free radicals are a mitigating factor in ischemia-reperfusion injury. They are believed to play a major role in postischemic dysfunction (stunning) associated with ischemic bouts of less than about 20 minutes.[6] Much of the injury is thought to result from the burst of free radicals during the initial moments of reperfusion. The relative contribution of free radicals associated with irreversible injury from more prolonged ischemic bouts is less clear. Henry et al.[47] have observed that the magnitude of free radical generation during reperfusion is dependent on ischemic duration in a bell-shaped manner: low after short and long lengths and highest at intermediate lengths. As reviewed by Marban and Bolli,[6] exogenous administration of superoxide dismutase (SOD) and catalase via the coronary circulation has proven to be effective in attenuating myocardial stunning although these enzymes cannot penetrate cells of the heart. Administration of these enzymes in this manner does not appear to be effective against prolonged ischemic insults that result in irreversible injury (infarction). However, elevations of intracellular antioxidants using cell-permeant antioxidants has been reported to attenuate reperfusion injury in animals subjected to 45 minutes, and even 90 minutes, of coronary occlusion.[48–50] Furthermore, transgenic mice overexpressing catalase or SOD have enhanced cardioprotection against ischemia-reperfusion injury.[51–53]

Superoxide dismutase is sometimes considered to be the first line of defense against intracellular oxygen-derived free radicals. It forms hydrogen peroxide from the superoxide radical and is located in both mitochondria (Mn-SOD) and the cytosol (CuZn-SOD). Catalase and the glutathione redox system work to prevent the accumulation of hydrogen peroxide. This molecule is an important source of oxygen free radical species, including the highly reactive hydroxyl radical, but it also acts independently and freely diffuses across biological membranes.[54] Although both catalase and the glutathione system act to metabolize hydrogen peroxide, there is considerable controversy concerning the relative roles of these two antioxidant systems in the myocardium under physiological and pathological conditions.

Up-regulation of antioxidant enzyme genes may play a significant role in cardioprotection following heat shock. Many of the studies reporting improved post-ischemic functional recovery within 24 hours after heat shock also reported increased catalase and/or SOD.[21,55–58] As mentioned, Karmazyn et al.[21] reported that the irreversible catalase inhibitor, 3-aminotriazole, abolished the protective effect of heat shock on postischemic mechanical recovery of isolated perfused rat hearts. Also, Joyeux et al.[56] reported that catalase plays a role in preventing reperfusion arrhythmias of heat stressed rats. In contrast, Auyeung et al.[59] reported that 3-aminotriazole does not alter the heat shock attenuation of infarct size subsequent to *in vivo* ischemia, and Steare and Yellon[57] reported that the increase in endogenous catalase activity caused by heat stress does not protect isolated rat heart against exogenous hydrogen peroxide. Yamashita et al.[58] found that heat stress to isolated rat cardiomyocytes caused the induction of Mn-SOD and HSP70 and was associated with improved tolerance to hypoxia 24 hours later. When the elevation of Mn-SOD was prevented by an antisense oligodeoxyribonucleotide to the enzyme, the enhanced tolerance to hypoxia following heat stress was also prevented, although HSP70 was increased as before. This finding is interesting because HSP70 alone can protect cells and isolated hearts from ischemia (as discussed above). Perhaps SOD and HSP70 can work in concert to facilitate the biological activity of each other.

Mitochondrial SOD is not generally considered to be the "weak link" responsible for the ischemia-reperfusion dysfunction associated with stunning.

As stated in a recent review by Bolli and Marban,[6] there is considerable evidence that the underlying mechanism(s) for myocardial stunning is due to disruptions in cytosol/sarcolemma. In fact, they dismiss mechanisms that involve mitochondria as "not plausible."[6] This conclusion appears to be consistent with our earlier experiments on hearts stunned by short-term ischemia.[60] In this study, we found that isolated mitochondrial function was unaltered while overall cardiac function was depressed. However, Mn-SOD may play an increasingly more prominent role as the duration or severity of the ischemic stress increases.

V. Exercise and cardioprotection

Exercise has been found to provide many of the same cardioprotective benefits as heat shock. Participation in a chronic exercise program has been reported to attenuate myocardial ischemia-reperfusion injury in both isolated heart models and *in vivo* models.[60–69] As in heat shock studies, cardioprotection has been observed 24 hours after acute exercise bouts in previously sedentary animals.[70–72] Although the mechanisms are unclear, many investigators have focused on increased expression of various cardioprotective proteins. Because exercise can result in the rapid induction of HSP70, it has been implicated as one of the proteins responsible for the intrinsic cardioprotection observed after acute exercise[70,71] and chronic exercise.[61,65,68] Thus, it appears that exercise and heat stress may share some common pathways in providing protection against ischemia-induced injury.

A.   Chronic exercise studies

Prior to 1990, epidemiological studies suggested that exercise could provide cardioprotection.[1,3,4] However, studies specifically investigating whether exercise provided intrinsic protective adaptations to the heart were split approximately 50:50 between yes and no (see Reference 73 for review). Much of the confusion was due to differences in exercise mode and ischemic models employed. Therefore, in 1992, we carried out the first study to determine whether treadmill running in rats can improve recovery of hemodynamic function in isolated hearts following global ischemia.[61] In this study, male Fischer 344 rats were treadmill-trained for an hour per day for 3 to 4 months at three intensities ranging from low (20 meters/min, 0% grade) to intense (30 meters/min with ten 2-minute sprints at 60 meters/min, 5% grade). Cardiac function at low and high workloads was evaluated before and after 25 minutes of global, zero-flow ischemia in the isolated, working heart model. Evaluation of cardiac performance prior to ischemia revealed that none of the treadmill training protocols affected basic intrinsic cardiac pump performance. However, after ischemia plus 45 minutes of subsequent reperfusion, cardiac function in all trained groups was significantly better than that of the sedentary rats (Figure 5.2). Furthermore, the magnitude of improvement tended to increase with exercise intensity. A non-ischemic

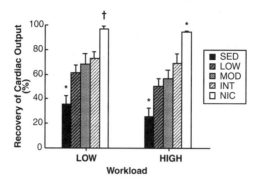

Figure 5.2 Effect of exercise training at various intensities on relative recovery of cardiac output after 25 minutes of global ischemia in isolated working hearts. Rats ran on a treadmill for 3 to 4 months for 1 hour per day at low (LOW), moderate (MOD), or high (INT) intensities. Values are means ± SE. SED, sedentary; LOW, 20 m/min, 0% grade; MOD, 30 m/min, 5% grade; INT, 30 m/min with ten 2-minute sprints at 60 m/min, 5% grade; NIC, non-ischemic control. *P < 0.05 vs. all other groups; †P < 0.05 vs. all groups except INT. (Modified from Bowles, D. K., Farrar, R. P., and Starnes, J. W., *Am. J. Physiol.,* 263, H804, 1992.)

control group of sedentary rats maintained 97.3% of cardiac output over this same time period, indicating that the preparation was quite stable and that almost all of the decreases in function were due to actual ischemia-induced cardiac dysfunction. The lack of an improvement in pre-ischemic cardiac performance in the exercise-trained rats indicates that improved postischemic recovery of cardiac function with training is not dependent on a training-enhanced pre-ischemic pump performance.

Other studies have since observed that many biochemical processes are also better preserved after an ischemic bout in chronically trained rats. Bowles and Starnes[60] found that the increased postischemic contractile function in trained hearts was associated with decreased calcium overload, a greater inotropic response to extracellular calcium, increased efficiency of cardiac work, and a lower diastolic stiffness. Although energy status was less in the sedentary group, isolated mitochondrial function was unaltered, which is consistent with a stunned myocardium. Demirel et al.[62] and Powers et al.[65] observed that better functional recovery was associated with less lipid peroxidation and oxidative stress following *in vivo* ischemia-reperfusion in endurance-trained rats compared to sedentary animals. These authors used a treadmill training program that was progressively increased to maintain a constant relative exercise intensity throughout the 10 weeks of exercise. During the last week of the program, the rats were running 90 minutes/day at 30 meters/minute up an 18% grade, which represents a very strenuous exercise bout. Although the better maintenance of calcium handling abilities and/or energy status and less oxidative stress in the postischemic exercise-trained hearts could be responsible for their better performance, the specific underlying adaptations that protect these systems was not explored.

It is apparent from Figure 5.2 that considerable cardioprotection can be achieved by relatively light exercise; however, the minimum required has not been established. Libonati et al.[64] reported no attenuation of reperfusion injury in Sprague-Dawley rats exercised on a treadmill for 1 hour/day, 5 days/week, for 6 weeks at a low intensity (20 m/min 0% grade). However, rats subjected to a sprint protocol were found to have less reperfusion injury than controls or low-intensity runners. The authors' interpretation is based solely on the finding that during recovery from a 20-minute global ischemia bout in Langendorff perfused hearts, the sprint-trained group had higher absolute values of developed pressure (LVDP) and rate of pressure development (dP/dt). However, if the authors had compared recovery to pre-ischemic function in each heart, their conclusion might have been different. It appears from the data presented that *both* exercise groups recovered 100% of LVDP and 125% of dP/dt compared to only 86% and 104%, respectively, in the control group. This high recovery by all groups, along with low enzyme leakage during reperfusion, indicates that the injury was a form of stunning; therefore, the Langendorff preparation may not have been sensitive enough to make a determination of possible differences in mechanical function. Interestingly, action potential duration measurements in myocytes isolated from these hearts indicated that calcium handling was better preserved after ischemia in both low-intensity endurance and sprint-trained groups compared to controls. Another factor that confounds the interpretation of this study is that all groups performed an endurance run to exhaustion and a graded intensity treadmill performance test at the end of the training period. These acute procedures alone may have influenced the ischemia-reperfusion response in all groups, thus masking potential differences (see Section V.B on acute exercise studies).

Voluntary running wheels, which are often used for exercise studies, do not appear to provide exercise of sufficient intensity to influence intrinsic cardioprotection. In a recent study, ad libitum-fed and calorie-restricted male Wistar rats were allowed free access to a running wheel for specific periods each day for 8 months.[74] The authors report that the number of revolutions per session was low in the ad libitum group and much higher in the calorie-restricted group. No biochemical or performance measure of training status was determined in either group. Isolated perfused working hearts were subjected to 25 minutes of global ischemia and subsequently reperfused for 15 minutes. A two-way factorial analysis was used to determine whether exercise or the interaction of calorie restriction and exercise had an effect on recovery of heart function. The statistical analysis revealed that recovery of aortic systolic pressure or aortic flow was not improved in either running group. However, the authors state that the calorie restricted runners may have had better recovery than the calorie restricted sedentary controls because aortic flow recovery tended to be higher ($66 \pm 7\%$ vs. $56 \pm 7\%$, mean $\pm$ standard error). Other measures that are typically evaluated in working heart preparations, such as intraventricular systolic/diastolic pressure, cardiac output and coronary flow, were not determined. Interestingly, Noble et al.[75] did not find HSP70 to increase in ad libitum-fed rats given access to voluntary running wheels.

At least one study employing the isolated perfused working heart preparation appears to be in disagreement regarding the effects of chronic exercise on post-ischemic functional recovery. Using a long-term, chronic exercise program, Paulson et al.[76] trained rats at 21 meters/min, 5% grade for 90 min/day. Isolated working hearts were evaluated before and after 75 minutes of low-flow ischemia. In this method of ischemia, perfusion of the coronary circulation is continued but at a severely restricted rate so that cardiac oxygen demand greatly exceeds oxygen supply. Training was found to have no effect on recovery of cardiac output or work; both groups declined approximately 25% compared to pre-ischemic values. However, during the low-flow ischemic period, the perfusate was supplemented with 22 mM glucose, which amounted to 297 grams of glucose being provided during ischemia. High levels of extracellular glucose are known to protect the myocardium from hypoxic damage[77] and extend the time to onset of contracture during ischemia.[78] In addition, their study lacked a non-ischemic control group to determine whether a decline in function occurred in the absence of ischemia due to the prolonged perfusion. Thus, a training-induced effect in the study of Paulson et al.[76] may have been obscured by the benefit of exogenous glucose to the control hearts and/or the lack of a time-matched normoxic control group. Recently, Margonato et al.[67] reported that isolated perfused Langendorff hearts from swim-trained rats had improved recovery of rate-pressure product and fewer arrhythmias than sedentary animals following 60 minutes of low-flow ischemia with perfusate supplemented with 11 mM glucose. Although the training protocols differed, the fact that Margonato et al. was able to differentiate between sedentary and exercised animals, using the relatively crude Langendorff preparation following an ischemic bout similar to that employed by Paulson's group, casts further doubt on their conclusions.

We recently completed a study to determine which of many potential cardioprotective proteins were relevant to cardioprotection associated with a chronic exercise program.[68] The concentration of several proteins were evaluated at 3, 6, and 9 weeks of a 9-week exercise program. During the last 6 weeks of the program, the treadmill running intensity (20 meters/min up a 6% grade) and duration (60 minutes) were kept constant so that we could follow the adaptive process. To investigate the effect of exercise independent of the associated increase in body temperature, some rats were run with wetted fur in an 8°C room to maintain body temperature at resting temperature. The proteins measured included those that are the primary focus of this chapter, that is, the inducible form of HSP70, catalase, cytosolic superoxide dismutase (CuZn-SOD), mitochondrial SOD (Mn-SOD), and glutathione peroxidase. We also measured other stress proteins, including glucose regulated protein 75 (GRP75 also known as mitochondrial HSP70), heme oygenase-1 (HO-1 also known as HSP32), HSP90, and αB-crystallin. Left ventricular contents of only three of these proteins changed over time in a manner

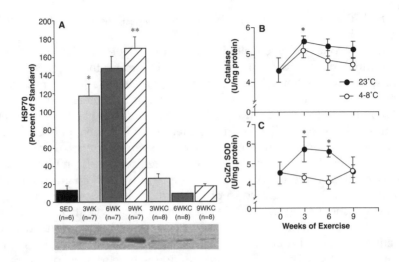

Figure 5.3   Left ventricular HSP70 (panel A), catalase (panel B), and CuZn-SOD (panel C) at 3, 6, and 9 weeks of exercise in a 23°C room (WK) and an 8°C room (WKC). Body temperature did not change from resting temperature when exercising in the colder room. Values are means ± SE. (From Harris, M. B. and Starnes, J. W., *Am. J. Physiol.*, 280, H2271, 2001. With permission.)

consistent with enhanced cardioprotection. HSP70 increased over time in rats whose body temperature increased during exercise but did not increase at any time during 9 weeks of daily exercise in rats whose body temperature was kept at resting temperature during exercise (Figure 5.3A). Catalase was increased in the early phase of the exercise program regardless of body temperature (Figure 5.3B). CuZn-SOD was significantly elevated at 3 and 6 weeks in the animals whose body temperature increased, but was not different from sedentary animals at any time during the training program in the animals whose body temperature was clamped at resting temperature (Figure 5.3C). Thus, changes in HSP70, catalase, and CuZn-SOD were prevented or attenuated when body temperature was held constant. After 9 weeks, all antioxidant enzymes were similar to control values regardless of exercise temperature. Consistent with other studies, mechanical recovery of isolated perfused working hearts following 22.5 minutes of global ischemia was enhanced after chronic training in the room-temperature environment, but the effect was abolished by 9 weeks of exercise in the cold environment. Overall, this study provides strong evidence that the primary cardioprotective protein induced by chronic exercise is HSP70. However, it should be pointed out that the cold running model resulted in cardiac hypertrophy (approximately 25% increase in mass), which may have decreased tolerance to ischemia-reperfusion, thus masking benefits accrued through exercise. Also, the intensity of exercise employed was low; perhaps a more intense training program would have induced a cardioprotective effect in the group exercised in the cold.

The finding by Harris and Starnes[68] that adaptation to a chronic exercise program resulted in protection against ischemia-reperfusion injury in the absence of increases in antioxidant enzymes is consistent with other chronic exercise studies.[62,64,69] Specifically, it was observed that some antioxidant enzymes in the ventricles were elevated in the initial stages of an exercise program, but eventually returned toward sedentary levels as the individual adapted to the imposed exercise load. However, it is important to point out that measurement of an enzyme activity in whole ventricle may be misleading. Failing cardiac myocytes have been reported to have greater suscept-ability to oxygen free radical-mediated injury although antioxidant enzyme levels are normal.[79] Specific changes in small compartments, (e.g., coronary vasculature) may have gone unnoticed in bulk measures of left ventricular contents. Rush et al.[80] have recently reported that cytosolic SOD is increased in pig coronary arterioles following 3 to 4 months of exercise on a motorized treadmill. As discussed later, changes in the vasculature could have impor-tant cardioprotective implications.

B.   Acute exercise studies

Locke et al.[70] were the first to discover that acute exercise results in enhanced postischemic myocardial recovery. In this study, rats were exercised on a motor-driven treadmill for 1 hour at a speed of 30 meters/minute up a 15% grade and sacrificed 24 hours after exercising once or for 3 consecutive days. Another group was heat shocked (15 minutes at 42°C) without exercising. Isolated hearts were perfused using the Langendorff preparation and sub-jected to 30 minutes of global ischemia and 30 minutes of reperfusion. Rats run for 3 days and those that were heat shocked had significantly improved postischemic recovery of left ventricular function compared to controls, but animals run for only 1 day were similar to controls. HSP70 levels were significantly elevated in the 3-day runners, but not in the 1-day runners, and catalase activity was increased by heat shock alone, but not by exercise. The authors concluded that the increased recovery in the exercising rats was related to HSP70 content and proposed that the mechanism for enhanced post-ischemic functional recovery is related, at least in part, to increased expression of HSP70.

In a subsequent study, Taylor et al.[71] confirmed and extended these findings. They evaluated postischemic functional recovery using the isolated perfused working rat heart from groups of animals treated similarly to those in the Lock et al. study. In addition, a group was run in a cold environment so that core temperature did not increase during exercise. Taylor et al. found that all groups had enhanced postischemic functional recovery 24 hours after acute exercise or heat treatment. Left ventricular HSP70 was elevated several-fold in the groups that were heat shocked or exercised at normal room temperature, but was not elevated above control level in the group exercised in the cold. When all exercise groups were combined, there was no significant correlation between HSP70 expression and myocardial reperfusion injury

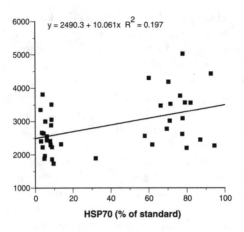

Figure 5.4   Relationship between myocardial HSP70 content and postischemic function of isolated working rat hearts after 1 to 3 days of acute exercise. CO × SP, cardiac output times systolic pressure. The lack of a significant correlation suggests that attenuated reperfusion injury after acute exercise may be partially due to elevated HSP70 and partially due to some other factor. (Modified from Taylor, R. P., Harris, M. B., and Starnes, J. W., *Am. J. Physiol.*, 276, H1098, 1999. With permission.)

following acute exercise in previously sedentary animals (Figure 5.4). The authors concluded that increased HSP70 expression following acute exercise is primarily due to increased core temperature and is not the only factor that can provide improved postischemic function.

Yamashita et al.[72] looked specifically at whether cardioprotection following acute exercise might be related to the activation of Mn-SOD. After 2 weeks of acclimating to treadmill running, male Wistar rats weighing 240 to 300 grams were exercised for 25 to 30 minutes at 27 to 30 meters/minute on a 0% grade. At various times post-exercise (from 30 minutes to 72 hours), *in vivo* ischemia was imposed by ligating the left coronary artery for 20 minutes. Myocardial infarct size was determined 48 hours after reperfusion. Exercise was found to decrease infarct size in a biphasic manner that coincided with the activity of Mn-SOD at the time of ischemia (Figure 5.5A). At 30 minutes (early phase) and 48 hours (late phase) post-exercise, Mn-SOD was approximately 65% above control and ischemia-induced infarct size was significantly decreased. At all other times (between 30 minutes and 48 hours, or after 48 hours), Mn-SOD activity and infarct size did not differ from sedentary. The transient increase in Mn-SOD activity at 30 minutes post-exercise was due to a post-translational modification of the enzyme, while the increase at 48 hours was due to an increase in its content. When the increase in Mn-SOD content was prevented by an antisense oligodeoxyribonucleotide, the decrease in infarct size during the late phase was abolished.

Yamashita et al.[72] also observed that the cytokines, tumor necrosis factor α(TNF-α) and interleukin 1β (IL-1β), were increased at the end of the exercise bout for a brief period. Preventing the exercise-induced elevation of both by

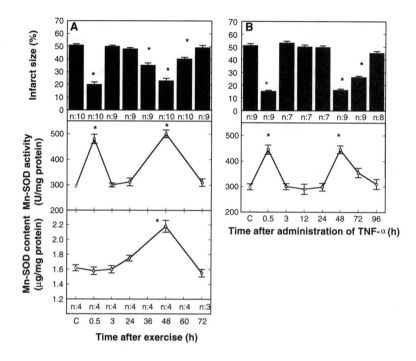

Figure 5.5 Relationship between Mn-SOD activity and postischemic infarct size at various times after a single exercise bout (panel A) or administration of TNF-α (panel B). Rats were subjected to *in vivo* coronary artery occlusion for 20 minutes at the times indicated, and the size of the infarct was evaluated 48 hours after reperfusion. Values are means ± SE. *P <0.05 vs. control group. (From Yamashita, N. et al., *J. Exp. Med.*, 189, 1699, 1999. With permission.)

administering neutralizing antibodies or the antioxidant MPG (which did not affect infarct size in sham-treated controls) abolished the biphasic pattern of infarct size and Mn-SOD activity. Administration of TNF-α to sedentary rats resulted in the same biphasic pattern of infarct size and Mn-SOD activity observed with exercise (Figure 5.5B). Thus, they proposed the following mechanism for cardioprotection associated with acute exercise. Reactive oxygen species generated during exercise cause increases in TNF-α and IL-1β, which activates Mn-SOD during the early phase and induces Mn-SOD expression during the late phase.

The data of Yamashita et al.[72] provide compelling evidence that exercise-induced Mn-SOD is important in protecting against irreversible myocardial injury; however, it may not protect against stunning. Functional measures, consisting of rate-pressure product, were not different among groups during 20 minutes of regional ischemia or 30 minutes of subsequent reperfusion. Although rate-pressure product provides only a crude indication of function, the results imply that elevation of mitochondrial SOD did not protect against stunning. This is consistent with the view of Marban and Bolli,[6] who say that stunning is the result of cytosolic/sarcolemmal injury and does not involve mitochondria. Yamashita et al.[72] report that the cytosolic form of SOD was not elevated by acute running and catalase activity was not measured. However, other investigators have found CuZn-SOD[81] and catalase[82] to be elevated immediately after exercise.

VI. Coronary vasculature dysfunction

It is now well-established from a variety of experimental models that the coronary vasculature can be a target of damage or dysfunction during ischemia-reperfusion.[83] When this occurs, it can contribute to myocardial ischemia-reperfusion injury.[84] Thus, agents or procedures that can preserve vasomotor function and integrity following an ischemia-reperfusion stress can enhance myocardial protection. Ischemic preconditioning has been found to protect against the loss of endothelium-dependent vasodilation associated with an ischemia-reperfusion bout 24 hours later.[85] The proposed stimulus for this cardioprotective adaptation was free radical generation during the preconditioning event.

Recently, Chen et al.[53] carried out an ischemia-reperfusion study using transgenic mice overexpressing cytosolic SOD (CuZn-SOD) only in coronary endothelial and smooth muscle cells. They found the transgenic animals to have better recovery of myocardial function and less enzyme leakage than wild-type controls after 35 minutes of ischemia and 45 minutes of reperfusion in a Langendorff perfusion model. There were no differences between transgenics and wild-types for levels of myocardial cell CuZn-SOD, other antioxidant enymes, HSP70, or HSP25. Chronic exercise training by Yucatan pigs appears to result in a similar phenotype. Rush et al.[80] found an increased content and activity of the cytosolic form of SOD in coronary arterioles without a change in Mn-SOD, catalase, or $p67^{phos}$, which is a component of the superoxide-generating NAD(P)H oxidase. The above findings have important ramifications under both normal and ischemia-reperfusion situations. As explained in Figure 5.6, increased superoxide scavenging capacity via CuZn-SOD in endothelial cells simultaneously enhances vasodilation by raising nitric oxide (NO) and H_2O_2 levels while attenuating the production of the damaging substances (peroxynitrite and hydroxyl radical).[86] The role of NO as a vasodilator is well known, while the finding that H_2O_2 directly hyperpolarizes smooth muscle cells, causing further dilation, is a more recent discovery. In addition to protecting against increased free radical production that occurs during reperfusion, enhancing the dilatory stimulus might help prevent vasoconstriction that can occur paradoxically with increased cardiac work in some coronary disease patients.[87]

There is very little information available regarding whether exercise attenuates coronary vasculature dysfunction associated with ischemia-reperfusion. Bowles et al.[61] observed that chronic exercise training resulted in higher coronary flow upon reperfusion of isolated rat hearts. However, vascular function was not specifically investigated. Recently, the first study (and to our

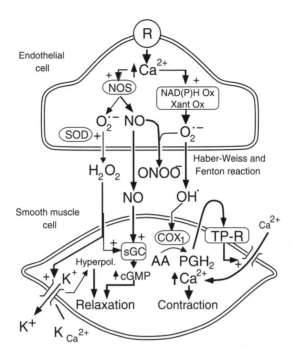

Figure 5.6   Proposed interactions between nitric oxide (NO) and superoxide anions ($O_2^{\cdot-}$) in the regulation of endothelium-dependent responses. NO synthase (NOS) produces both NO and superoxide anions. Under normal circumstances, and in most arteries, the production of NO predominates, and NO scavenges the small amounts of superoxide anion formed. Superoxide anions that escape the scavenging by NO are transformed by SOD to H_2O_2, which diffuses to the vascular smooth muscle and causes its hyperpolarization (Hyperpol.) by opening of a K^+ conductance (K_{Ca}^{2+}). NO activates soluble guanylate cyclase (sGC) to produce more cGMP. Other sources of production of superoxide anions (e.g., NAD(P)H oxidase [NAD(P)H Ox]) or xanthine oxidase (Xant Ox) are activated when the intracellular Ca^{2+} concentration increases. The large quantities of superoxide anions formed scavenge most or all of the NO, leading to the production of peroxynitrite (ONOO⁻). In addition, superoxide anions can be transformed to hydroxyl radicals, which diffuse to the vascular smooth muscle and induce the production of vasoconstrictor endoperoxides (PGH$_2$) and prostanoids (and possibly isoprostanes). The latter activate TP receptors (TP-R) that are coupled positively to the contractile process. (From Vanhoutte, P. M., *J. Clin. Invest.*, 107, 23, 2001. With permission.)

knowledge, the only study) evaluating the effect of chronic exercise on coronary vascular function following ischemia-reperfusion was carried out by Symons et al.[66] Female Sprague-Dawley rats ran in a 23°C room for 60 minutes/day at 26.8 meters/minute up a 15% grade for the last 5 to 7 weeks of a 10 to 12 week exercise program. Stress protein levels were not evaluated, but the exercise program as described should easily result in a several-fold increase in HSP70. Ischemia was produced by occluding the left coronary artery *in situ* for three 15-minute periods separated by two reperfusion periods of 5 minutes. Coronary resistance vessels, which are responsible for regulating myocardial blood flow, were isolated beginning 10 minutes after the last ischemia bout and their function subsequently evaluated *in vitro* using wire myographs. Following ischemia-reperfusion, vessels from trained animals developed less constriction in response to endothelin-1 and U-46619, which act on distinctly different receptor types to induce contraction, and to KCl which acts via voltage-gated Ca^{2+} channels (nonreceptor-mediated). However, relaxation was not different from sedentary animals in response to both endothelium-dependent and endothelium-independent vasodilators. The authors conclude that chronic training lessens receptor-mediated vaso-constriction but does not affect endothelial function of coronary resistance vessels after ischemia-reperfusion. Furthermore, the blunted contractile response was concluded to be due to postreceptor mechanisms affecting calcium regulation. In nonischemic time controls, the response to endothelin-1 was not different between trained and sedentary groups, indicating that the difference following ischemia was the result of better preservation of calcium regulation in the trained group. However, other studies have found a reduction in the contractile response of nonischemic vessels from exercise trained animals in response to agonists that mobilize sarcoplasmic reticulum Ca^{2+}.[88] In animals displaying this exercise adaptation, perhaps a greater degree of cardioprotection compared to sedentary animals would be observed.

In the paper by Symons et al.,[66] myocardial contractile function (e.g., developed pressure, dP/dt, and systolic wall thickening) did not differ between trained and sedentary animals during the *in situ* ischemia-reperfusion protocol employed. The reasons for this are unclear but could be partially due to the very short recovery time allowed (10 minutes) after reperfusion before determining mechanical function. Evidence supporting this possibility is that mechanical function following "long-term" ischemia (total of 45 minutes) was the same as that following a single, 5-minute, "short-term" ischemic bout, from which the heart would be expected to fully recover. Also, the experimental model employed (regional ischemia *in vivo*) is not the most sensitive model for comparing mechanical function. It should be pointed out, however, that Symons et al.[66] designed their study primarily to evaluate postischemic coronary vascular function, which was carried out at least an hour after reperfusion.

VII. Summary

There is now considerable evidence from both *in vivo* and *in vitro* experimental models that exercise training results in intrinsic adaptations that render the heart better able to recover from an ischemic bout. These adaptations include the increased expression of certain stress proteins. This exercise benefit can be realized within 1 or 2 days following an initial exercise bout of sufficient intensity and duration. Some of these proteins may be transient in nature and eventually return to sedentary levels as adaptation to the exercise program occurs, while at least one, the stress-induced member of the 70-kDalton heat shock protein family (HSP70), persists throughout an exercise program. The fact that HSP70 provides very effective protection has been established but the mechanism by which it protects has not. A recent report indicates that cytosolic superoxide dismutase is elevated in the coronary vasculature of chronically exercised pigs[80] and another indicates that coronary vasculature dysfunction after ischemia is attenuated in chronically exercised rats.[66] Whether the two observations are related remains to be determined but changes in the vasculature could have important cardioprotective implications. Finally, it appears that the exercise-related cardioprotection can be achieved with only moderate exercise but the minimum exercise load required has not been determined.

References

1. Morris, J. N. et al., Vigorous exercise in leisure-time: protection against coronary heart disease, *Lancet, 6*, 1207, 1980.
2. Mittleman, M. A. et al., Triggering of acute myocardial infarction by heavy physical exertion: protection against triggering by regular exercise, *New Engl. J. Med., 329,* 1677, 1993.
3. Paffenbarger, R. S., Jr., Wing, A. L., and Hyde, R. T., Physical activity as an index of heart attack risk in college alumni, *Am. J. Epidemiol., 108,* 161, 1978.
4. Rechnitzer, P. A. et al., Long-term follow-up study of survival and recurrence rates following myocardial infarction in exercising and control subjects, *Circulation, 45,* 853, 1972.
5. Braunwald, E. and Kloner, R. A., The stunned myocardium; prolonged, postischemic ventricular dyfunction, *Circulation, 66,* 1146, 1982.
6. Bolli, R. and Marban, E., Molecular and cellular mechanisms of myocardial stunning, *Physiolog. Rev., 79,* 609, 1999.
7. Homans, D. C. et al., Cumulative deterioration of myocardial function after repeated episodes of exercise-induced ischemia, *Am. J. Physiol., 256,* H1462, 1989.
8. Homans, D. C. et al., Persistence of regional left ventricular dysfunction after exercise-induced myocardial ischemia, *J. Clin. Invest., 77,* 66, 1986.
9. Hittinger, L. et al., Exercise-induced subendocardial dysfunction in dogs with left ventricular hypertrophy, *Circ. Res., 66,* 329, 1990.
10. Kehrer, J. P. and Starnes, J. W., Models and markers used to study cardiac reperfusion injury, *Pharmacol. Therapeut., 44,* 123, 1989.
11. Ytrehus, K., The ischemic heart — Experimental models, *Pharmacolog. Res., 42,* 193, 2000.
12. Langendorff, O., Untersuchungen am überlebenden Säugetierherzen, *Pflügers Arch., 61,* 291, 1895.
13. Arnold, G. et al., The importance of the perfusion pressure in the coronary arteries for the contractility and the oxygen consumption of the heart, *Pflügers Arch., 299,* 339, 1968.
14. Vogel, W. M. et al., Acute alterations in left ventricular diastolic chamber stiffness. Role of the "erectile" effect of coronary arterial pressure and flow in normal and damaged hearts, *Circ. Res., 51,* 465, 1982.

15. Neely, J. R. et al., Effect of pressure development on oxygen consumption by isolated rat heart, *Am. J. Physiol.,* 225, 651, 1967.
16. Ritossa, F., A new puffing pattern induced by temperature shock and DNP in *Drosophila, Experentia,* 18, 571, 1962.
17. Tissières, A., Mitchell, H. K., and Tracy, U. M., Protein synthesis in salivary glands of *Drosophila melanogaster*: relation to chromosome puffs, *J. Mol. Biol.,* 84, 389, 1974.
18. Li, G. C. and Hahn, G. M., Ethanol-induced tolerance to heat and to adriamycin, *Nature,* 274, 699, 1978.
19. Currie, R. W. et al., Heat-shock response is associated with enhanced postischemic ventricular recovery, *Circ. Res.,* 63, 543, 1988.
20. Currie, R. W. and Karmazyn, M., Improved post-ischemic ventricular recovery in the absence of changes in energy metabolism in working rat hearts following heat-shock, *J. Mol. Cell. Cardiol.,* 22, 631, 1990.
21. Karmazyn, M., Mailer, K., and Currie, R. W., Acquisition and decay of heat-shock-enhanced postischemic ventricular recovery, *Am. J. Physiol.,* 259, H424, 1990.
22. Marber, M. S. et al., Attenuation by heat stress of a submaximal calcium paradox in the rabbit heart, *J. Mol. Cell. Cardiol.,* 25, 1119, 1993.
23. Marber, M. S. et al., Myocardial protection after whole body heat stress in the rabbit is dependent on metabolic substrate and is related to the amount of the inducible 70-kD heat stress protein, *J. Clin. Invest.,* 93, 1087, 1994.
24. Yellon, D. M. et al., The protective role of heat stress in the ischaemic and reperfused rabbit myocardium, *J. Mol. Cell. Cardiol.,* 24, 895, 1992.
25. Donnelly, T. J. et al., Heat shock protein induction in rat hearts. A role for improved myocardial salvage after ischemia and reperfusion?, *Circulation,* 85, 769, 1992.
26. Qian, Y. Z. et al., Dissociation of heat shock proteins expression with ischemic tolerance by whole body hyperthermia in rat heart, *J. Mol. Cell. Cardiol.,* 30, 1163, 1998.
27. Walker, D. M. et al., Heat stress limits infarct size in the isolated perfused rabbit heart, *Cardiovasc. Res.,* 27, 962, 1993.
28. Yellon, D. M. et al., Whole body heat stress fails to limit infarct size in the reperfused rabbit heart, *Cardiovasc. Res.,* 26, 342, 1992.
29. Mocanu, M. M. et al., Heat stress attenuates free radical release in the isolated perfused rat heart, *Free Radic. Biol. Med.,* 15, 459, 1993.
30. Fink, A. L., Chaperone-mediated protein folding, *Physiolog. Rev.,* 79, 425, 1999.
31. Lindquist, S. and Craig, E. A., The heat-shock proteins, *Annu. Rev. Genetics,* 22, 631, 1988.
32. Benjamin, I. J. and McMillan, D. R., Stress (heat shock) proteins: molecular chaperones in cardiovascular biology and disease, *Circ. Res.,* 83, 117, 1998.
33. Hutter, M. M. et al., Heat-shock protein induction in rat hearts. A direct correlation between the amount of heat-shock protein induced and the degree of myocardial protection, *Circulation,* 89, 355, 1994.
34. Hutter, J. J. et al., Overexpression of heat shock protein 72 in transgenic mice decreases infarct size *in vivo, Circulation,* 94, 1408, 1996.
35. Marber, M. S. et al., Overexpression of the rat inducible 70-kD heat stress protein in a transgenic mouse increases the resistance of the heart to ischemic injury, *J. Clin. Invest.,* 95, 1446, 1995.
36. Plumier, J. C. et al., Transgenic mice expressing the human heat shock protein 70 have improved post-ischemic myocardial recovery, *J. Clin. Invest.,* 95, 1854, 1995.

37. Radford, N. B. et al., Cardioprotective effects of 70-kDa heat shock protein in transgenic mice, *Proc. Natl. Acad. Sci. U.S.A.*, 93, 2339, 1996.

38. Trost, S. U. et al., Protection against myocardial dysfunction after a brief ischemic period in transgenic mice expressing inducible heatshock protein 70, *J. Clin. Invest.*, 101, 855, 1998.

39. Suzuki, K. et al., *In vivo* gene transfection with heat shock protein 70 enhances myocardial tolerance to ischemia-reperfusion injury in rat, *J. Clin. Invest.*, 99, 1645, 1997.

40. Suzuki, K. et al., Reduction in myocardial apoptosis associated with overexpression of heat shock protein 70, *Basic Res. Cardiol.*, 95, 397, 2000.

41. Kawana, Ki. et al., Cytoprotective mechanism of heat shock protein 70 against hypoxia/reoxygenation injury, *J. Mol. Cell. Cardiol.*, 32, 2229, 2000.

42. Knowlton, A. A., Mutation of amino acids 566-572 (KKKVLDK) inhibits nuclear accumulation of heat shock protein 72 after heat shock, *J. Mol. Cell. Cardiol.*, 33, 49, 2001.

43. Sakaguchi, T. et al., Ecto-5′-nucleotidase plays a role in the cardioprotective effects of heat shock protein 72 in ischemia-reperfusion injury in rat hearts, *Cardiovasc. Res.*, 47, 74, 2000.

44. Joyeux, M. et al., Protein kinase C is involved in resistance to myocardial infarction induced by heat stress, *J. Mol. Cell. Cardiol.*, 29, 3311, 1997.

45. Kukreja, R. C. et al., Role of protein kinase C and 72 kDa heat shock protein in ischemic tolerance following heat stress in the rat heart, *Mol. Cell. Biochem.*, 195, 123, 1999.

46. Yamashita, N. et al., Time course of tolerance to ischemia-reperfusion injury and induction of heat shock protein 72 by heat stress in the rat heart, *J. Mol. Cell. Cardiol.*, 29, 2815, 1997.

47. Henry, T. D. et al., Postischemic oxygen radical production varies with duration of ischemia, *Am. J. Physiol.*, 264, H1478, 1993.

48. Forman, M. B. et al., Glutathione redox pathway and reperfusion injury. Effect of N-acetylcysteine on infarct size and ventricular function, *Circulation*, 78, 202, 1988.

49. Klein, H. H. et al., The effects of Trolox, a water-soluble vitamin E analogue, in regionally ischemic, reperfused porcine hearts, *Int. J. Cardiol.*, 32, 291, 1991.

50. Puett, D. W. et al., Oxypurinol limits myocardial stunning but does not reduce infarct size after reperfusion, *Circulation*, 76, 678, 1987.

51. Li, G. et al., Catalase-overexpressing transgenic mouse heart is resistant to ischemia-reperfusion injury, *Am. J. Physiol.*, 273, H1090, 1997.

52. Chen, Z. et al., Overexpression of MnSOD protects against myocardial ischemia/reperfusion injury in transgenic mice, *J. Mol. Cell. Cardiol.*, 30, 2281, 1998.

53. Chen, Z. et al., Overexpression of CuZnSOD in coronary vascular cells attenuates myocardial ischemia/reperfusion injury, *Free Radic. Biol. Med.*, 29, 589, 2000.

54. Halliwell, B., Antioxidant defense mechanisms: from the beginning to the end (of the beginning), *Free Radic. Res.*, 31, 261, 1999.

55. Liu, X. et al., Heat shock: a new approach for myocardial preservation in cardiac surgery, *Circulation*, 86, 358, 1992.

56. Joyeux, M. et al., *In vitro* antiarrhythmic effect of prior whole body hyperthermia: implication of catalase, *J. Mol. Cell. Cardiol.*, 29, 3285, 1997.

57. Steare, S. E. and Yellon, D. M., Increased endogenous catalase activity caused by heat stress does not protect isolated rat heart against exogenous hydrogen peroxide, *Circ. Res.*, 28, 1096, 1994.

58. Yamashita, N. et al., Heat shock-induced manganese superoxide dismutase enhances the tolerance of cardiac myocytes to hypoxia-reoxygenation injury, *J. Mol. Cell. Cardiol.*, 29, 1805, 1997.

59. Auyeung, T. et al., Catalase inhibition with 3-amino-1,2,4-triazole does not abolish infarct size reduction in heat-shocked rats, *Circulation*, 92, 3318, 1995.

60. Bowles, D. K. and Starnes, J. W., Exercise training improves metabolic response after ischemia in isolated working rat heart, *J. Appl. Physiol.*, 76, 1608, 1994.

61. Bowles, D. K., Farrar, R. P., and Starnes, J. W., Exercise training improves cardiac function after ischemia in the isolated, working rat heart, *Am. J. Physiol.*, 263, H804, 1992.

62. Demirel, H. A. et al., Exercise training reduces myocardial lipid peroxidation following short-term ischemia-reperfusion, *Med. Sci. Sports Exerc.*, 30, 1211, 1998.

63. Ji, L. L. et al., Cardiac hypertrophy alters myocardial response to ischaemia and reperfusion *in vivo*, *Acta Physiol. Scand.*, 151, 279, 1994.

64. Libonati, J. R. et al., Reduced ischemia and reperfusion injury following exercise training, *Med. Sci. Sports Exerc.*, 29, 509, 1997.

65. Powers, S. K. et al., Exercise training improves myocardial tolerance to *in vivo* ischemia-reperfusion in the rat, *Am. J. Physiol.*, 275, R1468, 1998.

66. Symons, J. D. et al., Microvascular and myocardial responses to ischemia: influence of exercise training, *J. Appl. Physiol.*, 88, 433, 2000.

67. Margonato, V. et al., Swim training improves myocardial resistance to ischemia in rats, *Int. J. Sports Med.*, 21, 163, 2000.

68. Harris, M. B. and Starnes, J. W., Effects of body temperature during exercise training on myocardial adaptations, *Am. J. Physiol.*, 280, H2271, 2001.

69. Kihlstrom, M., Protection effect of endurance training against reoxygenation-induced injuries in rat heart, *J. Appl. Physiol.*, 68, 1672, 1990.

70. Locke, M. et al., Enhanced postischemic myocardial recovery following exercise induction of HSP 72, *Am. J. Physiol.*, 269, H320, 1995.

71. Taylor, R. P., Harris, M. B., and Starnes, J. W., Acute exercise can improve cardioprotection without increasing heat-shock protein, *Am. J. Physiol.*, 276, H1098, 1999.

72. Yamashita, N. et al., Exercise provides direct biphasic cardioprotection via manganese superoxide dismutase activation, *J. Exp. Med.*, 189, 1699, 1999.

73. Starnes, J. W. and Bowles, D. K., Role of exercise in the cause and prevention of cardiac dysfunction, *Exerc. Sport Sci. Rev.*, 23, 349, 1995.

74. Broderick, T. L. et al., Effects of chronic food restriction and exercise training on the recovery of cardiac function following ischemia, *J. Gerontol.: Biol. Sci.*, 56A, B33, 2001.

75. Noble, E. G. et al., Differential expression of stress proteins in rat myocardium after free wheel or treadmill run training, *J. Appl. Physiol.*, 86, 1696, 1999.

76. Paulson, D. J. et al., Improved postischemic recovery of cardiac pump function in exercised trained diabetic rats, *J. Appl. Physiol.*, 65, 187, 1988.

77. Barry, A. C., Barry, G. D., and Zimmerman, J. A., Protective effect of glucose on the anoxic myocardium of old and young mice, *Mechanism Ageing Dev.*, 40, 41, 1987.

78. Owen, P., Dennis, S., and Opie, L. H., Glucose flux regulates onset of ischemic contracture in globally underperfused rat hearts, *Circ. Res.*, 66, 344, 1990.

79. Tsutsui, H. et al., Greater susceptibility of failing cardiac myocytes to oxygen free radical-mediated injury, *Cardiovasc. Res.*, 49, 103, 2001.

80. Rush, J. W. E. et al., SOD-1 expression in pig coronary arterioles is increased by exercise training, *Am. J. Physiol.*, 279, H2068, 2000.
81. Radak, Z. et al., Superoxide dismutase derivative reduces oxidative damage in skeletal muscle of rats during exhaustive exercise, *J. Appl. Physiol.*, 79, 129, 1995.
82. Somani, S. M., Frank, S., and Rybak, L. P., Responses of antioxidant system to acute and trained exercise in rat heart subcellular fractions, *Pharmacol. Biochem. and Behav.*, 51, 627, 1995.
83. Rubino, A. and Yellon, D. M., Ischaemic preconditioning of the vasculature: an overlooked phenomenon for protecting the heart?, *Trends Pharmacolog. Sci.*, 21, 225, 2000.
84. Qi, X. L. et al., Vascular endothelial dysfunction contributes to myocardial depression in ischemia-reperfusion in the rat, *Can. J. Physiol. Pharmacol.*, 76, 35, 1998.
85. Kaeffer, N., Richard, V., and Thuillez, C., Delayed coronary endothelial protection 24 hours after preconditioning: role of free radicals, *Circulation*, 96, 2311, 1997.
86. Vanhoutte, P. M., Endothelium-derived free radicals: for worse and for better, *J. Clin. Invest.*, 107, 23, 2001.
87. Sambuceti, G. et al., Coronary vasoconstriction during myocardial ischemia induced by rises in metabolic demand in patients with coronary artery disease, *Circulation*, 95, 2652, 1997.
88. Bowles, D. K., Laughlin, M. H., and Sturek, M., Exercise training alters the Ca^{2+} and contractile response of coronary arteries to endothelin, *J. Appl. Physiol.*, 78, 1079, 1995.

chapter six

Heat shock proteins and reactive oxygen species

Karyn L. Hamilton and Scott K. Powers

Contents

I. Introduction

Induction of heat shock proteins (HSPs) is one of the best-described cellular responses to stress.[1-6] Indeed, many different stresses such as heat stress, hypoxia, acidosis, exposure to oxidants, and ischemia/reperfusion have been shown to increase the expression of these highly conserved proteins.[1,3,7-15] As reviewed in Chapter 3, the physiological stress associated with endurance exercise training also results in a rapid induction of several isoforms of HSPs; this induction of cellular HSPs is associated with cellular protection from a variety of stresses (e.g., heat and oxidant damage).[16-26] To date, the exact cellular

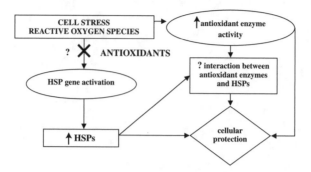

Figure 6.1   Proposed relationship between HSPs and ROS. Cell stress, including stress mediated by ROS, leads to up-regulation of HSPs and antioxidant enzyme activity and ultimately to protection against subsequent cell stress. Exogenous anti-oxidants may interfere with activation of HSP genes, thus inhibiting HSP synthesis and related cell protection. HSPs may or may not exert their cellular protective effects in concert with antioxidant enzymes.

signal responsible for the exercise-associated induction of HSPs remains unclear. Similarly, the mechanism(s) by which HSPs provide protection against subsequent cellular stresses is incompletely understood.

   This chapter reviews our current knowledge about the relationship between reactive oxygen species (ROS) and HSPs. We will begin with a brief review of ROS, cellular oxidative injury, and antioxidant defense mechanisms available to the cell. A discussion of the evidence for increased production of ROS serving as a signal for up-regulation of HSPs will follow. Finally, we conclude with a discussion of HSPs and protection against ROS-mediated cellular damage (see Figure 6.1).

II. Reactive oxygen species and oxidative stress: an overview

ROS such as the peroxyl radical (ROO'), the superoxide anion ($O_2^{.-}$), and the hydroxyl radical (HO') are derived from the univalent reduction of molecular oxygen.[28] Certain ROS are known as free radicals because they contain one or more unpaired electrons in their outer orbital (e.g., ROO', $O_2^{.-}$, HO'). When a ground-state O_2 molecule accepts a single electron, the product is $O_2^{.-}$ and addition of a second electron yields hydrogen peroxide (H_2O_2). Homolytic fission of the O–O bond in H_2O_2 produces two HO' molecules and can be achieved by a simple mixture of H_2O_2 and an iron (ferrous) salt. Called the Fenton reaction, the HO' produced by this fission is capable of reacting at very high rate constants with almost every cellular component.[29]

A. Sources of ROS in skeletal and cardiac muscle

Potential sources of ROS in skeletal muscle fibers during exercise include the mitochondrial respiratory chain, xanthine oxidase production of superoxide, enzymatic arachadonic acid oxygenation, nitric oxide synthesis, catecholamine oxidation, and neutrophil-induced oxidative bursts.[30] The relative contributions of each of these pathways during exercise remain an active area of research.

It is well-established that ROS are produced during ischemia and reperfusion (I-R) in both skeletal and cardiac muscles.[31,32] A potential source of oxidants during both ischemia and reperfusion is the univalent reduction of oxygen in complexes I and III of the mitochondrial electron transport chain.[31] The enzymatic oxidation by xanthine oxidase of purines such as xanthine and hypoxanthine is another potential source of ROS during I-R in rodents.[33,34] Primarily located in the vessel walls of most tissues, including cardiac muscle, the enzyme xanthine dehydrogenase (XDH) catalyzes the oxidation of hypoxanthine to xanthine, and xanthine to uric acid. During I-R, XDH may be either reversibly or irreversibly transformed to xanthine oxidase. In contrast to its dehydrogenase form, xanthine oxidase utilizes O_2 as the electron acceptor and produces O_2^- while catalyzing the oxidation of hypoxanthine to uric acid.[35] In postischemic tissues, activated neutrophils are known to release large amounts of ROS that have also been associated with injury.[36] The relative contribution of any of these sources of ROS in the induction of HSPs associated with exercise remains undefined.

B. ROS-mediated cellular damage

The cellular toxicity associated with ROS is largely reflective of their rapid participation in damaging reaction with cellular components.[37] The damage resulting from these reactions includes lipid peroxidation, protein oxidation, and DNA damage.[36] During hypoxia-reoxygenation injury, for example, cellular proteins, membrane lipids, and DNA are particularly vulnerable to oxidative damage.[37–41] Damaged proteins and by-products of phospholipid-bilayer degradation may serve as important triggers for HSP gene induction.[4,18–19,42–47] A brief overview of oxidative damage to cellular lipids and proteins follows.

1. Lipid peroxidation and protein oxidation

Polyunsaturated fatty acids (PUFAs) are highly susceptible to ROS attack and oxidative damage at their sites of unsaturation. Briefly, the attack begins when ROS have sufficient energy to abstract hydrogen atoms from the methylene groups of PUFAs; this forms lipid peroxyl radicals that can then react with other PUFAs, starting an amplifying chain reaction.[48] Peroxyl radicals are also capable of propagating further oxidative damage to other membrane-associated molecules such as proteins. The overall result of this attack is destruction of membrane PUFAs leading to alteration of membrane fluidity and permeability. By-products of lipid peroxidation, such as malondialdehyde, lipid hydroperoxides, conjugated dienes, and isoprostanes, are often measured to estimate the extent of ROS-mediated tissue damage.[28] Furthermore, release of arachadonic acid, a sensitive marker of membrane phospholipid degradation, has been assayed as a measurement of membrane damage resulting from ROS-mediated injury such as that seen during postischemic reperfusion.[44,46]

Included in the ROS-mediated damage to cellular proteins is oxidation of contractile, cytoskeletal, and membrane-bound proteins and enzymes.[39,48] Quantitative studies of protein oxidation reveal that proteins containing sulfhydryl groups are highly susceptible to damage.[48] Much of the damage caused by protein oxidation is irreparable, and proteins that have been damaged by ROS are susceptible to further proteolytic cleavage. Oxidation of proteins results in formation of carbonyl groups and assays to determine protein carbonyl content have been used as an estimation of oxidative damage to proteins.[49]

The combined effects of lipid peroxidation and protein oxidation have a profound and detrimental effect on cell structure and function.[50] However, cells are armed with intrinsic mechanisms designed to defend against oxidative damage. The following section highlights key components of cellular antioxidant defenses.

III. Antioxidant defense mechanisms

Cells have extensive endogenous and exogenous mechanisms for providing protection against damage by ROS. Among the cells' primary defenses are both enzymatic antioxidants and nonenzymatic antioxidants. The enzymatic antioxidants include (1) superoxide dismutase, which dismutates the superoxide radical; (2) catalase, which decomposes hydrogen peroxide; and (3) glutathione peroxidase, which decomposes hydrogen peroxide with the assistance of the nonenzymatic antioxidant glutathione.[28,51] Among the nonenzymatic antioxidants are vitamin E, vitamin C, carotenoids, glutathione, ubiquinones, and flavonoids. Antioxidants work both independently and synergistically to maintain a reduced cellular environment.[51,52]

Given the potential role of ROS in cellular injury, interventions designed to increase the cells' antioxidant defenses are a promising strategy for minimizing cell damage such as that associated with hypoxia-reoxygenation. While increases in cellular enzymatic antioxidants can be achieved following stresses including exercise training,[52,53] another effective method for increasing antioxidant defenses is via dietary supplementation with nonenzymatic antioxidants. Indeed, enhanced antioxidant defenses have been reported to prevent ROS-associated damage to cellular components.[54–65] However, the effect of enhancing antioxidant capacity on the cells' ability to express HSPs remains unclear. Literature addressing this issue is discussed in the following sections.

IV. Transcription of HSP genes

As reviewed in Chapters 1 and 2 of this book, a wide variety of physical and chemical changes evoking cellular stress are associated with synthesis of HSPs in tissues.[4,18–19,43,47] Despite the fact that transduction pathways involved in the activation of HSP genes have been actively investigated, the mechanisms underlying these pathways remain controversial, especially with respect to the "primary sensor" at the level of the cell membrane. In contrast, the mechanisms involved in the intracellular signaling cascade leading to activation of HSP gene transcription and protein synthesis are better understood. In eukaryotic cells, transcription is orchestrated by transcription factors, referred to as heat shock factors (HSFs), that are inactive in unstressed cells. Cell stress leads to HSF activation, DNA binding, and ultimately gene transcription.[66] Four HSFs have been identified thus far; HSF1 is the primary stress response heat shock factor in mammalian cells. The reader is referred to several excellent reviews for a detailed description of the regulation and activation of HSF1.[4,18–19,43,47]

A. ROS and activation of HSP genes

Few biological molecules have a negative reputation equal to ROS. Indeed, ROS have gained the reputation as injurious molecules because of their capability of promoting cellular damage and their contribution in the pathogenesis of a variety of diseases. Despite this reputation as injurious molecules, recent evidence indicates that ROS are important messengers in cellular signal transduction and transcription factor activation.[67–71] In addition to regulating pivotal transcription factors such as nuclear factor-κB (NF-κB) and activator protein-1 (AP-1), ROS have also been reported by many groups to result in HSF1 activation and HSP gene induction (see Table 6.1).[72–78] For example, Nishizawa et al.[78] reported significant activation of HSF1 and appearance of HSP70 and 90 mRNA with both ischemia-reperfusion and with H_2O_2 perfusion in isolated hearts. Similarly, perfusion of hearts with xanthine plus xanthine oxidase (for generation of the superoxide radical), H_2O_2, and illuminated rose bengal (for generation of singlet oxygen) all resulted in significant increases in HSP70 mRNA.[76] Interestingly, it has been suggested that one pathway through which ROS activates HSF1 is via oxidation of protein thiols. Nonnative disulfide bond formation, induced by chemicals such as N-ethyl-maleimide or menadione redox cycling, have been shown to result in protein denaturation, HSF1 activation, and accumulation of HSP70 mRNA and protein.[45,73]

B. Antioxidants and activation of HSP genes

Given the evidence that ROS can serve as a stimulus for expression of HSPs, it follows that enhancing cellular antioxidant defenses could inhibit HSP transcription, resulting in an attenuation of HSP accumulation (see Table 6.2). One of the first experiments to test this notion employed both heat shock

Table 6.1 Impact of Reactive Oxygen Species and Other Types of Stress on Cellular Expression of Heat Shock Proteins

Cell Stress	Effect on HSP Species	Ref.
Ischemia-reperfusion (review)	Many species reviewed	72
Nonnative disulfide bond formation/protein denaturation induced by chemicals such as N-ethyl-maleimide or menadione redox cycling	HSF1 activation, and accumulation of HSP70 mRNA and protein	45,73
IL-1α	Increased HSP70	74
IL-1α and TNF	Increased small HSP mRNA	75
Xanthine plus xanthine oxidase (for generation of the superoxide radical), H_2O_2, and illuminated rose bengal (for generation of singlet oxygen)	Increased HSP70 mRNA	76
IL-1α	Increased HSP27	77
Ischemia-reperfusion and H_2O_2 perfusion	HSF1 activation and appearance of HSP70 and 90 mRNA	78

Table 6.2 Impact of Antioxidants and Cellular Stress on Expression of Heat Shock Proteins

Cell Stress	Effect of Antioxidants on HSPs	Ref.
Heat shock and hypoxia-reoxygenation	No change in any HSP	79
Reperfusion	No change in HSP70	80
Heat shock	Inhibited or attenuated HSP72; inhibited or attenuated HSP27	81
Erythrophagocytosis (in phagocytic cells)	Inhibited HSP32 and HSP70	82
Heat shock	Attenuated HSP70	84
Exercise	Attenuated HSP72	85
Hypoxia	Attenuated HSP32	86
Tobacco smoke	Inhibited HSP32 but not HSP70, 90, or 110	87
Ischemia-reperfusion	Inhibited HSF1 activation and attenuated accumulation of HSP70 and 90 mRNA	78
Ultraviolet A radiation	No change in HSP72	88

and hypoxia-reoxygenation to induce oxidant stress in *Drosophila*.[79] These experiments revealed that while both stresses resulted in accumulation of HSPs, only hypoxia-reoxygenation resulted in oxidant stress as measured by oxidized/reduced glutathione. Further, these experiments revealed that HSP accumulation was not affected by antioxidant treatment.[79] The role of

ROS generated during reperfusion in the induction of the HSP70 gene was further investigated using the antioxidants allopurinol and dimethyl sulfoxide.[80] Neither antioxidant, individually or in combination, changed the reperfusion-associated gene induction. These two experiments suggest that some antioxidants do not impair the expression of HSP70 in cells.

In a study investigating the effects of a variety of antioxidant compounds on heat shock response of cultured myelocytic cells, pyrrolidine dithiocarbamate and 1,10-phenanthroline inhibited hsp72 accumulation and N-acetylcysteine attenuated it.[81] Pyrrolidine dithiocarbamate and 1,10-phenanthroline also inhibited hsp27 while N-acetylcysteine and allopurinol decreased its accumulation. Kantengwa and Polla[82] studied the roles of oxygen radicals in HSP induction by erythrophagocytosis in phagocytic cells and reported an inhibition of HSP32 and HSP70 induction by a flavonoid antioxidant. To further evaluate this mechanism, the authors also employed an inhibitor of protein kinase C (PKC) which abolished the effect of antioxidants on induction of all HSPs studied except heme oxygenase (HSP32). PKC has subsequently been shown to play a pivotal role in regulation of HSP72 gene activation.[83] Collectively, these studies support the notion that different HSP species may be more sensitive to modulation of cellular redox status and that, specifically, the induction of "classic" HSPs and heme oxygenase is differentially regulated.

Ushakova et al.[84] pursued the role of ROS in the heat shock response utilizing a whole-body hyperthermia model to stress mice fed an antioxidant-supplemented diet. The authors reported that in normothermic mice, dietary supplementation either had no influence on HSP70 gene expression (spleen) or enhanced it (brain, liver). Antioxidant supplementation attenuated HSP70 gene expression following heat shock in all tissues.[84] In agreement with this, we have recently observed that antioxidant supplementation attenuates exercise-associated HSP72 accumulation.[85] Interestingly, this antioxidant-associated attenuation of HSP72 did not result in a parallel decrease in protection against ischemia-reperfusion injury.

Accumulation of HSPs following environmental stresses other than hyperthermia and exercise has also been shown to be attenuated by antioxidant exposure. N-acetylcysteine, a thiol donor shown to enhance reduced glutathione concentrations, attenuated HSP32 gene induction following hypoxia.[86] Using tobacco smoke as a stress, Pinot et al.[87] demonstrated that both quercetin (a flavonoid) and N-acetylcysteine prevented the induction of heme oxygenase (HSP32) but not that of HSP70, 90, or 110. In contrast, ischemia-reperfusion associated HSF1 activation and accumulation of HSP70 and 90 mRNA were decreased by exposure of isolated perfused hearts to allopurinol (a xanthine oxidase inhibitor) or catalase.[78] Expression of HSP72 induced by ultraviolet-A radiation was not, however, impacted by exposure to vitamin E and butylated hydroxytoluene in a fibrosarcoma cell line.[88] Hence, it appears that the effects of antioxidants on HSP gene expression vary with the type of cell stress, the stress protein isoform, the amount and type of antioxidants, and the cell type under investigation.

V. HSPs and protection against ROS-mediated injury

While the functions of stress proteins are reviewed extensively in other chapters, it is appropriate to briefly review the existing evidence for a role of HSPs in protecting cells from damages mediated by ROS. Clearly, evidence suggests that many of the same cellular stresses that result in activation of HSP genes, also up-regulate antioxidant enzyme activity. For example, several groups have demonstrated that induction of HSPs correlates with increases in the activity of catalase, an enzyme that decomposes H_2O_2.[89–92] Reports of modification of other antioxidant enzyme activities, such as glutathione peroxidase and superoxide dismutase, in correlation with HSP induction are varied, with some investigations indicating that high levels of HSPs augment antioxidant enzyme activity whereas others report that HSPs do not influence antioxidant enzyme activity.[89,92–98] Therefore, at present, the role, if any, that stress proteins play in enhancing the functions of cellular antioxidants remains unclear. Nonetheless, by stimulating synthesis or by preventing aggregation or degradation of oxidized cellular proteins, HSPs would complement existing antioxidants during and following cellular oxidative stress. This remains an important and exciting area for future research.

VI. Summary

This chapter has provided an overview of the relationship between ROS and HSPs. In summary, numerous ROS are produced in both skeletal and cardiac muscle cells. Although the exact site of production for these ROS continues to be investigated, evidence exists that ROS are produced in the mitochondrial respiratory chain, xanthine oxidase pathway, arachadonic acid metabolism, nitric oxide production, and oxidation of catecholamines. The cellular toxicity associated with ROS comes from the participation of ROS in damaging reactions in cells. In particular, ROS result in damage to proteins (protein oxidation), lipid oxidation (lipid peroxidation), and DNA damage. Numerous enzymatic and nonenzymatic antioxidant defense mechanisms exist in the cell; collectively, these defense mechanisms work to reduce oxidative stress.

Growing evidence suggests a strong link between ROS and HSPs; this relationship is simplistically illustrated in Figure 6.1. First, it is well-established that ROS can trigger the activation of HSP genes. Further, it seems likely that antioxidants can inhibit the ROS-mediated activation of HSP gene transcription. Finally, evidence indicates that HSPs may provide protection and/or recovery from ROS-mediated injury. Improving our knowledge of the relationship between redox regulation and HSP expression and action is an important area for future research.

Acknowledgments

This work was supported, in part, by grants from the American Heart Association–Florida and the National Institutes of Health (NHLBI RO1 HL62361) awarded to S.K.P.

References

1. Ellis, R. J., Stress proteins as molecular chaperones, in *Stress Proteins in Medicine,* van Eden, W. and Young, D. B., Eds., Marcel Decker, New York, 1996, 1–26.
2. Lindquist, S., The heat shock response, *Annu. Rev. Biochem.,* 55, 1151, 1986.
3. Lindquist, S. and Craig, E. A., The heat-shock proteins, *Annu. Rev. Genet.,* 22, 631, 1988.
4. Morimoto, R. I. et al., The heat-shock response: regulation and function of heat-shock proteins and molecular chaperones, *Essays Biochem.,* 32, 17, 1997.
5. Ritossa, F. M., A new puffing pattern induced by a temperature shock and DNP in *Drosophilia, Experientia,* 18, 571, 1962.
6. Tissières, A., Mitchell, H. K., and Tracy, U. M., Protein synthesis in salivary glands of *Drosophilia melanogaster, J. Mol. Biol.,* 84, 389, 1974.
7. Hightower, L. E., Cultured animal cells exposed to amino acid analogues or puromycin rapidly synthesize several polypeptides, *J. Cell. Physiol.,* 102, 407, 1980.
8. Levinson, W., Oppermann, H., and Jackson, J., Transition series metals and sulf-hydryl reagents induce the synthesis of four proteins in eukaryotic cells, *Biochim. Biophys. Acta,* 606, 170, 1980.
9. Mestril, R. et al., Isolation of a novel inducible rat heat shock protein (HSP70) gene and its expression during ischemia-hypoxia and heat shock, *Biochem. J.,* 298, 561, 1994.
10. Mestril, R. et al., Expression of inducible stress protein 70 in rat heart myogenic cells confers protection against stimulated ischemia induced injury, *J. Clin. Invest.,* 93, 759, 1994.
11. Mestril, R. and Dillmann, W. H., Heat shock proteins and protection against myocardial ischemia, *J. Mol. Cell. Cardiol.,* 27, 45, 1995.
12. Welch, H. et al., HSP 90 chaperones protein folding *in vitro, Nature,* 358, 169, 1992.
13. Welch, W. J., The mammalian stress response: cell physiology, and biochemistry of stress proteins, in *Stress Proteins in Biology and Medicine,* Morimoto, R. I., Tissières, A., and Georgopopulus, C., Eds., Cold Spring Harbor Laboratory Press, Cold Spring Harbor, NY, 1990, 223.
14. Welch, W. J., Mammalian stress response: cell physiology, structure/function of stress proteins, and implications for medicine and disease, *Physiol. Rev.,* 72, 1063, 1992.
15. Welch, W. J. et al., Response of mammalian cells to metabolic stress; changes in cell physiology and structure/function of stress proteins, *Curr. Top. Microbiol. Immunol.,* 167, 31, 1991.
16. Fehrenbach, E. and Niess, A. M., Role of heat shock proteins in the exercise response, *Exerc. Immunol. Rev.,* 5, 57, 1999.
17. Kregel, K. C. and Moseley, P. L., Differential effects of exercise and heat stress on liver. HSP70 accumulation with aging, *J. Appl. Physiol.,* 80, 547, 1996.
18. Locke, M., The cellular stress response to exercise: role of stress proteins, *Exerc. Sport Sci. Rev.,* 25, 105, 1997.
19. Locke, M. and Noble, E. G., Stress proteins: the exercise response, *Can. J. Appl. Physiol.,* 20(2), 155, 1995.
20. Locke, M., Noble, E. G., and Atkinson, B. G., Exercising mammals synthesize stress proteins, *Am. J. Physiol.,* 258, C723, 1990.
21. Locke, M. et al., Activation of heat-shock transcription factor in rat heart after heat shock and exercise, *Am. J. Physiol.,* 268, C1387, 1995.

22. Locke, M. et al., Enhanced postischemic myocardial recovery following exercise induction of HSP 72, *Am. J. Physiol.*, 269, H320, 1995.
23. Losowsky, M. S., Kelleher, J., and Walker, B. E., Intake and absorption of tocopherol, *Ann. N.Y. Acad. Sci.*, 202, 212, 1972.
24. Powers, S. K. et al., Exercise training improves myocardial tolerance to *in vivo* ischemia-reperfusion in the rat, *Am. J. Physiol.*, 275, R1468, 1998.
25. Salo, D. C., Donovan, C. M., and Davies, K. J., HSP70 and other possible heat shock or oxidative stress proteins are induced in skeletal muscle, heart, and liver during exercise, *Free Radic. Biol. Med.*, 11, 239, 1991.
26. Taylor R. P., Harris, M. B., and Starnes, J. W., Acute exercise can improve cardioprotection without increasing heat shock protein content, *Am. J. Physiol.*, 276, H1098, 1999.
27. Noble, E. G. et al., Differential expression of stress proteins in rat myocardium after free wheel or treadmill run training, *J. Appl. Physiol.*, 86, 1696, 1999.
28. Halliwell, B. and Gutteridge, J., *Free Radicals in Biology and Medicine*, Oxford University Press, New York, 1999.
29. Lloyd, R. V., Hanna, P. M., and Mason, R. P., The origin of the hydroxyl radical oxygen in the fenton reaction, *Free Radic. Biol. Med.*, 22, 885, 1997.
30. Janero, D., Therapeutic potential of vitamin E against myocardial ischemic-reperfusion injury, *Free Radic. Biol. Med.*, 10, 315, 1991.
31. Becker, L. B. et al., Generation of superoxide in cardiomyocytes during ischemia before reperfusion, *Am. J. Physiol.*, 277, H2240, 1999.
32. Park, Y., Kanekal, S., and Kehrer, J. P., Oxidative changes in hypoxic rat heart tissue, *Am. J. Physiol.*, 260, H1395, 1991.
33. Chambers, D. L. and Downey, J. M., Xanthine oxidase, free radicals, and myocardial ischemia, in *Handbook of Free Radicals and Antioxidants in Biomedicine*, Miguel, J., Quintilla, A. T., and Webereds, H., Eds., CRC Press, Boca Raton, FL, 1989, 263–273.
34. Downey, J., Free radicals and their involvement during long-term myocardial ischemia and reperfusion, *Annu. Rev. Physiol.*, 52, 487, 1990.
35. Hellsten, Y., The role of xanthine oxidase in exercise, in *Exercise and Oxygen Toxicity*, Sen, C. K., Packer, L., and Hanninen, O., Eds., Elsevier Science, Amsterdam, 1995, 211–234.
36. Serbinova, E. et al., Thioctic acid protects against ischemia-reperfusion injury in the isolated perfused Langerdorff heart, *Free Radic. Res. Commun.*, 17, 49, 1992.
37. Halliwell, B. and Chirico, S., Lipid peroxidation: its mechanism, measurement, and significance, *Am. J. Clin. Nutr.*, 57, 715S, 1993.
38. Haddock, P. S., Shattock, M. J., and Hearse, D. J., Modulation of cardiac Na^+-K^+ pump current: role of protein and nonprotein sulfhydryl redox status, *Am. J. Physiol.*, 269, H297, 1995.
39. Hein, S., Sceffold, T., and Schaper, J., Ischemia induces early changes to cytoskeletal and contractile proteins in diseased human myocardium, *J. Thoracic Cardiovasc. Surg.*, 110, 89, 1995.
40. Romaschin, A. D. et al., Conjugated dienes in ischemic and reperfused myocardium: an *in vivo* chemical signature of oxygen free radical mediated injury, *J. Mol. Cell. Cardiol.*, 19, 289, 1987.
41. Romaschin, A. D. et al., Subcellular distribution of peroxidized lipids in myocardial reperfusion injury, *Am. J. Physiol.*, 259, H116, 1990.
42. Anathan, J., Goldberg, A. L., and Voellmy, R., Abnormal proteins serve as eukaryotic stress signals and trigger the activation of heat shock genes, *Science*, 232, 522, 1986.

43. Benjamin, I. J. and McMillan, D. R., Stress (heat shock) proteins: molecular chaperones in cardiovascular biology and disease, *Circ. Res.*, 83, 117, 1998.
44. Jurivich, D. A. et al., Arachidonate is a potent modulator of human heat shock gene transcription, *Proc. Natl. Acad. Sci., U.S.A.*, 91, 2280, 1994.
45. McDuffee, A. T. et al., Proteins containing non-native disulfide bonds generated by oxidative stress can act as signals for the induction of the heat shock response, *J. Cell. Physiol.*, 171, 143, 1997.
46. van der Vusse, G. J. et al., Heat stress pretreatment mitigates postischemic arachadonic acid accumulation in rat heart, *Mol. Cell. Biochem.*, 85, 205, 1998.
47. Wu, C., Heat shock transcription factors: structure and regulation, *Annu. Rev. Cell Dev. Biol.*, 11, 441, 1995.
48. Alessio, H. M., Lipid peroxidation processes in healthy and diseased models, in *Exercise and Oxygen Toxicity*, Sen, C. K., Packer, L., and Hanninen, O., Eds., Elsevier Science, Amsterdam, 1994, 269–295.
49. Reznick, A. and Packer, L., Oxidative damage to proteins: spectrophotometric method for oxidatively modified proteins, *Meth. Enzymol.*, 223, 357, 1994.
50. Swies, J., Omogbai, E. K. I., and Smith, G. M., Occlusion and reperfusion induced arrhythmias in rats: involvement of platelets and effects of calcium antagonists, *J. Cardiovasc. Pharmacol.*, 15, 816, 1990.
51. Yu, B., Cellular defenses against damage from reactive oxygen species, *Physiolog. Rev.*, 74, 139, 1994.
52. Powers S. K. and Hamilton, K. L., Antioxidants and exercise, *Clinics Sports Med.*, 18, 525, 1999.
53. Powers, S. et al., Rigorous exercise training increases superoxide dismutase activity in the ventricular myocardium, *Am. J. Physiol.*, 265, H2094, 1993.
54. Barsacchi, R. et al., Increased ultra weak chemiluminescence from rat heart at postischemic reoxygenation: protective role of vitamin E, *Free Radic. Biol. Med.*, 6, 573, 1989.
55. Ferrari, R. et al., Vitamin E and the heart: possible role as an antioxidant, *Acta Vitaminol. Enzymol.*, 5, 11, 1983.
56. Frolov, V. A. and Kapustin, V. A., Effect of vitamin A and E on the contractile function of the heart in experimental myocardial infarction, *Kardiologiia*, 23, 93, 1983.
57. Fuenmayor, A. J. et al., Vitamin E and ventricular fibrillation threshold in myocardial ischemia, *Jap. Circ. J.*, 53, 1229, 1989.
58. Guanieri, C., Flamigni, F., and Caldarera, C., Subcellular localization of alpha-tocopherol and its effect on RNA synthesis in perfused rabbit heart, *Ital. J. Biochem.*, 29, 176, 1980.
59. Guanieri, C. et al., Effect of alpha-tocopherol on hypoxic-perfused and reoxygenated rabbit heart muscle, *J. Mol. Cell. Cardiol.*, 10, 893, 1978.
60. Klein, H. et al., Combined treatment with vitamins E and C in experimental myocardial infarction in pigs, *Am. Heart J.*, 118, 667, 1989.
61. Kotegawa, M. et al., Effect of alpha-tocopherol on high energy phosphate metabolite levels in rat heart by 31P-NMR using a Langerdorff perfusion technique, *J. Mol. Cell. Cardiol.*, 25, 1067, 1993.
62. Massey, K. D. and Burton, K. P., Alpha-tocopherol attenuates myocardial membrane-related alterations resulting from ischemia and reperfusion, *Am. J. Physiol.*, 256, H1192, 1989.
63. Meerson, F. Z. and Ustinova, E. E., Prevention of stress injury to the heart and its hypoxic contracture by using the natural antioxidant alpha-tocopherol, *Kardiologicia*, 22, 89, 1982.

64. Pyke, D. D. and Chan, A. C., Effects of vitamin E on prostacyclin release and lipid composition of the ischemic rat heart, *Arch. Biochem. Biophys.*, 277, 429, 1990.

65. Scholich, H., Murphy, M., and Sies, H., Antioxidant activity of dihydrolipoate against microsomal lipid peroxidation and its dependence on alpha-tocopherol, *Biochim. Biophys. Acta*, 1001, 256, 1989.

66. Morimoto, R. I., Cells in stress: transcriptional activation of heat shock genes, *Science*, 259, 1409, 1993.

67. Jackson, M. J., McArdle, A., and McArdle, F., Antioxidant micronutrients and gene expression, *Proc. Nutr. Soc.*, 57, 301, 1998.

68. Sen, C. K. and Packer, L., Antioxidant and redox regulation of gene transcription, *FASEB J.*, 10, 709, 1996.

69. Freeman, M. L. et al., Characterization of a signal generated by oxidation of protein thiols that activates the heat shock transcription factor, *J. Cell. Physiol.*, 164, 356, 1995.

70. Suzuki, Y. J., Forman, H. J., and Sevanian, A., Oxidants as stimulators of signal transduction, *Free Radic. Biol. Med.*, 22, 269, 1997.

71. Suzuki, Y. J. and Ford, G. D., Redox regulation of signal transduction in cardiac and smooth muscle, *J. Mol. Cell. Cardiol.*, 31, 345, 1999.

72. Das, D. K., Maulik, N., and Moraru, I. I., Gene expression in acute myocardial stress. Induction by hypoxia, ischemia, reperfusion, hyperthermia, and oxidative stress, *J. Mol. Cell. Cardiol.*, 27, 181, 1995.

73. Freeman, M. L. et al., Characterization of a signal generated by oxidation of protein thiols that activates the heat shock transcription factor, *J. Cell. Physiol.*, 164, 356, 1995.

74. Jornot, L., Mirault, M. E., and Junod, A. F., Differential expression of HSP70 stress proteins in human endothelial cells exposed to heat shock and hydrogen peroxide, *Am. J. Respir. Cell. Mol. Biol.*, 5, 265, 1991.

75. Kaur, P., Welch, W. J., and Saklatvala, J., Interleukin I and tumour necrosis factor increase phosphorylation of the small heat shock protein, *FEBS Lett.*, 258, 269, 1989.

76. Kukreja, R. C., Kontos, M. C., and Hess, M. L., Free radicals and heat shock protein in the heart, *Ann. N.Y. Acad. Sci.*, 793, 108, 1996.

77. Maulik, N. et al., Interleukin-1α preconditioning reduces myocardial ischemia reperfusion injury, *Circulation*, 88, 387, 1993.

78. Nishizawa, J. et al., Reactive oxygen species play an important role in the activation of heat shock factor 1 in ischemic-reperfused heart, *Circulation*, 99, 934, 1999.

79. Drummond, I. A. and Steinhardt, R. A., The role of oxidative stress in the induction of *Drosophila* heat-shock proteins, *Exp. Cell. Res.*, 173, 439, 1987.

80. Bardella, L. and Comolli, R., Differential expression of c-jun, c-fos and HSP70 mRNAs after folic acid and ischemia-reperfusion injury: effect of antioxidant treatment, *Exp. Nephrol.*, 2, 158, 1994.

81. Gorman, A. M. et al., Antioxidant-mediated inhibition of the heat shock response leads to apoptosis, *Fed. Eur. Biochem. Soc. Lett.*, 445, 98, 1999.

82. Kantengwa, S. and Polla, B. S., Flavonoids, but not protein kinase C inhibitors, prevent stress protein synthesis during erythrophagocytosis, *Biochem. Biophys. Res. Commun.*, 180, 308, 1991.

83. Yamashita, N. et al., Time-course of tolerance to ischemia-reperfusion injury and induction of heat shock protein 72 by heat stress in the rat heart, *J. Mol. Cell. Cardiol.*, 29, 1815, 1997.

84. Ushakova, T. et al., The effect of dietary supplements on gene expression in mice tissues, *Free Radic. Biol. Med.,* 20, 279, 1996.

85. Hamilton, K. L. et al., Protective strategies against myocardial ischemia reperfusion injury: exercise and antioxidants, *Free Radic. Biol. Med.,* in review.

86. Borger, D. R. and Essig, D. A., Induction of HSP32 gene in hypoxic cardiomyocytes is attenuated by treatment with N-acetyl-L-cysteine, *Am. J. Physiol.,* 274, H965, 1998.

87. Pinot, F. et al., Induction of stress proteins by tobacco smoke in human monocytes: modulation by antioxidants, *Cell Stress Chaperones,* 2, 156, 1997.

88. Trautinger, F. et al., Expression of the 72-kD heat shock protein is induced by ultraviolet A radiation in a human fibrosarcoma cell line, *Exp. Dermatol.,* 8, 187, 1999.

89. Currie, R. W. et al., Heat-shock response is associated with enhanced postischemic ventricular recovery, *Circ. Res.,* 63, 543, 1988.

90. Karmazyn, M., Mailer, K., and Currie, R. W., Acquisition and decay of heat-shock-enhanced postischemic ventricular recovery, *Am. J. Physiol.,* 259, H424, 1990.

91. Liu, X. et al., Heat shock. A new approach for myocardial preservation in cardiac surgery, *Circulation,* 86, II358, 1992.

92. Mocanu, M. M. et al., Heat stress attenuates free radical release in the isolated perfused rat heart, *Free Radic. Biol., Med.,* 15, 459, 1993.

93. Begonia, G. and Salin, M., Elevation of superoxide dismutase in *Halobacterium halobium* by heat shock, *J. Bacteriol.,* 173, 5582, 1991.

94. Brown, J. M. et al., Endotoxin pretreatment increases endogenous myocardial catalase activity and decreases reperfusion injury of isolated rat hearts, *Proc. Natl. Acad. Sci., U.S.A.,* 86, 2516, 1989.

95. Maulik, N. et al., Myocardial adaptation to ischemia by oxidative stress induced by endotoxin, *Am. J. Physiol.,* 269, C907, 1995.

96. Amrani, M. et al., Role of catalase and heat shock protein on recovery of cardiac endothelial and mechanical function after ischemia, *Cardioscience,* 4, 193, 1993.

97. Auyeung, Y. et al., Catalase inhibition with 3-amino-1,2,4-triazole does not abolish infarct size reduction in heat-shocked rats, *Circulation,* 92, 3318, 1995.

98. Steare, S. E. and Yellon, D. M., The protective effect of heat stress against reperfusion arrhythmias in the rat, *J. Mol. Cell. Cardiol.,* 25, 1471, 1993.

chapter seven

Stress proteins and exercise-induced muscle damage

Anne McArdle and Malcolm J. Jackson

Contents

I. Introduction

Skeletal muscle damage occurs following excessive or unaccustomed exercise. The mechanisms by which this damage occurs are relatively well-established. However, skeletal muscle can adapt rapidly such that the muscle is significantly protected against subsequent periods of exercise. This adaptation has been well-characterized at the structural level, but the cellular and biochemical changes that occur are less well-understood. The most comprehensively studied of these endogenous protective mechanisms in skeletal muscle are heat shock proteins (HSPs) and antioxidant enzymes. The aim of this review is to summarize evidence that skeletal muscle can produce HSPs, the mechanisms involved in this production, the effect of an increased content of HSPs in skeletal muscle on the susceptibility to damage. The role of an aberrant production of HSPs in muscle during the aging process will also be discussed.

II. Location and function of HSPs in resting skeletal muscle

Skeletal muscle consists of a heterogenous mix of cell or fiber types. These fibers differ in their functional properties and this is reflected in their HSP content. The fiber type of muscles has important implications on both the level of HSP content in unstressed fibers and the extent of increase in HSP content following stress. Data suggests that the nonstressed muscle comprising of primarily type 1 fibers contains a significantly higher amount of HSPs in comparison with barely detectable levels in a muscle that is predominantly composed of type 2 fibers.[1-3] Primarily, oxidative muscles have increased oxygen-derived radical formation during exercise and the increased baseline levels of HSPs may suggest a role for HSPs in providing protection against oxidative stress.

HSPs are present at low levels in unstressed skeletal muscle, where they act as molecular chaperones, associating with newly synthesized proteins and ensuring that the newly synthesized protein is folded and functions correctly. All HSPs act to preserve cellular integrity; HSP70 has the capacity to refold a wide array of proteins and other HSPs have more specific roles within the cell. The HSP70 family of proteins is by far the most studied in skeletal muscle. This family contains a constitutively expressed HSP of molecular mass 73 kDa, (known as HSC70 or HSP73) and a highly inducible HSP (known as HSP70 or HSP72),[4] as well as the glucose-related proteins (GRP75 and GRP78). HSC70 is constitutively expressed in the muscle cytoplasm and migrates to the nucleus during stress.[5] In contrast, HSP70 is present at low levels, with the exception of type 1 fibers of skeletal muscle,[1] in the cytoplasm of unstressed muscle. GRP75 and GRP78 are located in the mitochondria and the sarcoplasmic reticulum, respectively. GRP75 is thought to be involved in the translocation of newly synthesized proteins into the mitochondria and, in association with other HSPs, subsequent folding of the newly imported protein into a functioning state. GRP78 is thought to play a role in the assembly of secretory protein complexes.[6] HSP60 is primarily located within the muscle mitochondrial matrix, a potential key site of damage to muscle during exercise and aging. HSP60 and its co-chaperone (HSP10) constitute the chaperonin subclass of molecular chaperones, which facilitate folding of newly synthesized and imported proteins in mitochondria and are thought to be able to rescue proteins that denature spontaneously within the mitochondrion.[7,8] This is particularly relevant to skeletal muscle that contains a relatively high proportion of mitochondria and the activities of which respond rapidly to the increased energy demands of exercise. In contrast, αB-crystallin is a small HSP that is thought to play a key role in maintenance of cytoskeletal integrity in cells by association with various cytoskeletal proteins[9] and in muscle fibers is associated with the Z bands.[10] αB-Crystallin shares sequence homology with HSP25/27,[11] which is also present in the cytosol of skeletal muscle. HSP25/27 is thought to be involved in the stabilization of microfilaments. In addition, HSP25/27 can be phosphorylated and thus may play a role in cellular signal transduction.[12]

III. Exercise-induced muscle adaptation: the production of HSPs

Skeletal muscle is extremely responsive to changes in external stimuli such as neural stimuli, workload, or stretch. This training effect of skeletal muscle results in subsequent protection against damaging stresses and includes numerous structural adaptations and changes in gene expression, which result in biochemical modifications.[13] The most studied of these responses are the changes that occur in antioxidant enzymes and HSPs in muscle following a period of exercise. The production of HSPs in skeletal muscle of rodents following exercise has been relatively well-characterized, although the signal responsible for this adaptation is not clearly understood.

A. Production of HSPs in skeletal muscle following an acute period of exercise

Several workers have described increases in HSP mRNA and protein content following a period of mild or exhaustive exercise in rodents. Early studies examining the effect of treadmill running in rats for 20 minutes up to exhaustion demonstrated new or enhanced synthesis of proteins that corresponded to the Mr and pI of HSPs.[14] Subsequent studies showed increases in HSP expression following acute exhaustive exercise in rats.[15] Neufer et al.[3] demonstrated that induction of the HSP70 gene is evident within 24 hours of continuous low-frequency motor nerve stimulation of rabbit anterior tibialis muscle. Studies by this group have also demonstrated transient but large (>tenfold) increases in αB-crystallin and hsp70 mRNA following a single 8-hour bout of low-frequency (10 Hz) nerve stimulation of rabbit anterior tibialis muscles, and elevated levels of HSP60 and HSP70 protein in muscles after 14 and 21 days of continuous activation.[3,16] Studies from our laboratory have demonstrated that a 15-minute period of mild, nondamaging isometric contractions results in rapid increases in the production of HSPs in soleus and EDL muscles of mice.[2] In a similar manner to other studies, these data demonstrated that, in general, the level of HSPs in resting muscles composed

predominantly of type 1 fibers is higher than that of muscles that are predominantly type 2. In addition, the production of some HSPs appears to be
greater following exercise in the soleus muscle in comparison with the EDL
muscle. The most striking difference was observed in HSP60 measurements.
In soleus muscles of young Balb/c mice, the HSP60 content peaked at approximately 1500% of pre-exercised values at 18 hours following exercise,
whereas the HSP60 content of EDL muscles reached values of approximately
300% of resting values.[2]

B. Production of HSPs in skeletal muscle following chronic exercise

Various workers have examined the effect of endurance training on HSP
content of muscles. Ornatsky et al.[17] demonstrated an increase in HSP60 and
GRP75 and HSP70 content of anterior tibialis muscles following chronic
contraction at a frequency of 10 Hz for 10 days, and Ecochard et al.[18] demonstrated that muscles of rats that were sacrificed following up to 8 weeks
of endurance training displayed significantly increased HSP72 levels in skeletal muscles. Similarly, Gonzalez et al.[19] demonstrated an increased content
of HSP70, GRP75, and GRP78 in soleus muscles of rats subjected to an incre-
mental program of treadmill running of 3 months duration. In this study,
muscles of trained but rested animals experienced a stress response following acute exercise of lower intensity than that of the training sessions and
an inverse correlation was seen between HSP level and the rate of synthesis
of HSP72 during rest periods. These authors suggest that this is due to the
operation of a feedback regulatory loop aimed at reestablishing the threshold
levels characteristic of unstressed fibers. Studies by Samelman et al.[20] have
demonstrated that longer-term exercise in rats subjected to 16 to 20 weeks
of treadmill running, which resulted in an increase in mitochondrial biogenesis, is associated with an increased HSP60 content.

C. Production of HSPs in the skeletal muscle of humans

Exercise studies in humans have demonstrated that this effect also occurs in
human skeletal muscle. Puntschart et al.[21] showed that a period of exercise
in untrained subjects at their individual anaerobic threshold for 30 minutes
on a treadmill resulted in a significantly increased post-exercise content of
HSP70 mRNA. Similarly high levels were also observed 30 minutes and
3 hours after the end of exercise. Studies by Febbraio and Koukoulas[22] demonstrated that HSP mRNA increased significantly in human skeletal muscle
during prolonged exhaustive cycling exercise. However, Puntschart et al.[21]
did not detect a rise in HSP70 protein concentration within 3 hours after
cessation of exercise. This inability to detect changes in protein levels for up
to 3 hours following the exercise protocol can be explained by work from
our laboratory that demonstrated that HSP60 and HSP70 content of vastus
lateralis muscle was not significantly elevated until 3 to 6 days following a

period of exhaustive, nondamaging aerobic exercise in untrained subjects using one-legged cycle ergometry.[23] Interestingly, the variability in HSP70 content was remarkably high and evaluation of individual data illustrated that some subjects showed a clear rise in muscle HSP70 content within 24 to 48 hours following the exercise protocol, and others showed a slower and reduced proportionate rise in HSP70. This difference between subjects seemed to reflect the level of HSP70 in the muscle prior to exercise, where a muscle with a relatively low content of HSP70 responded to exercise by a rapid and large increase in HSP70 content, whereas a muscle with a relatively high content of HSP70 did not respond as rapidly to the exercise protocol.[23] In addition, Liu et al.[24] demonstrated that the HSP70 content of muscles of highly trained athletes increased during the training regime, with the maximum production at the end of the second week of training.

IV. Mechanisms responsible for activation of the stress response in skeletal muscle during exercise

Various physiological and chemical activators of the stress response have been identified in other cells. These include hyperthermia, oxidative stress, or incorporation of amino acid analogues into native proteins.[5] Skeletal muscle temperature can increase significantly during certain forms of exercise,[25] and Skidmore et al.[26] have examined the possibility that this change in temperature is responsible for activation of the stress response. In this study, muscles of rats were subjected to a prolonged bout of submaximal exercise in which colonic temperature remained at control levels. The exercise consisted of treadmill running at cooled, normal, and heated ambient temperatures for approximately 60 minutes. There were significant effects of both heating and exercise for HSP70 levels in the gastrocnemius and soleus muscles, although the increased HSP70 level during exercise was independent of core body temperature, suggesting that factors other than heat contribute to the increased expression of HSP70 during exercise.

Oxidative phosphorylation and formation of ATP are essential functions of mitochondria. Approximately 90% of cellular oxygen is metabolized within mitochondria. The univalent reduction of oxygen that takes place in mammalian tissues, including skeletal muscle, occurs at a rate of about 2% of the total oxygen uptake as a normal by-product of the electron transport chain[27] and this results in the production of superoxide radicals.[28] The oxygen consumption during isometric contractions (contractions in which the muscle is maintained at the same length during activation) or shortening (also known as miometric) contractions is also high and so the tendency to form free radicals during muscle contraction is high.[2,29] Thus, mitochondria, as a major site of ROS generation as well as a primary target of ROS in the cell, are thought to be particularly important in the age-related deterioration in muscle function.[30] There is considerable evidence to suggest that skeletal muscle produces an increased amount of free radicals during exercise. Data are conflicting, but several research groups have demonstrated that skeletal muscle generates a number of free radical species during contraction, including superoxide,[2,31] hydroxyl radical,[32] and nitric oxide.[33] In addition, Ohno et al.[34] concluded that levels of superoxide dismutase were generally elevated in muscles of young subjects after acute or chronic exercise, and similar data have been reported for changes in catalase activity.[35-37] There is a considerable amount of literature that has examined the possibility that this increased production of free radicals during exercise results is detrimental and leads to pathological changes within the muscle. However, recent studies have examined the possibility that the increased production of free radicals may act as a signal for activation of the stress response.[38] Studies in our laboratory and others have demonstrated that a period of muscle contractions results in the increased production of superoxide radicals[2] and hydroxyl radicals (see Reference 31 and unpublished observations). Previous data have shown that this pattern of stimulation induces a rise in the free radical signal seen on electron spin resonance examination of skeletal muscle.[39] This increased production of free radicals is accompanied by a transient oxidation of muscle protein thiols which reverses within 1 to 2 hours following the end of the contraction protocol.[2] Despite the oxidation of muscle proteins, no evidence of overt cellular damage was evident. Oxidation of muscle protein thiols has been shown to be part of a signaling mechanism leading to the induction of HSP expression in other cell types[40] and our findings in skeletal muscle appear to be compatible with this. Further evidence for a role of oxidative stress in the mechanism of induction of the stress response comes from supplementation studies carried out in our laboratory. Data demonstrated that prior supplementation with vitamin C, vitamin E, or β-carotene resulted in the attenuation or abolishment of the production of HSPs following a period of exercise that had previously been shown to activate the stress response.[41,42] The following scheme of activation events has been described.[38,43] In the unstressed cell, HSF1 is weakly associated with HSP70. Stress leads to the production of increased levels of misfolded or oxidized proteins and HSP70 has a higher affinity for binding these destabilized proteins. This sequestration of HSP70 releases HSF1 from its inactive monomeric state, allowing translocation to the nucleus, trimerization, hyperphosphorylation, and binding to the heat shock element (HSE) of HSP genes.

V. Skeletal muscle adapts to reduce the possibility of damage due to oxidative stress following contractile activity

Damage to skeletal muscle occurs following unaccustomed or excessive exercise, upon reperfusion following a period of ischemia, or as a consequence of muscle disorders such as Duchenne muscular dystrophy.[44] Although the initial insult can be widespread, muscle damage occurs by one of relatively few mechanisms.[44] Skeletal muscle has the unique ability to adapt rapidly following a period of damaging or nondamaging exercise, such that the muscle is considerably protected against subsequent (normally damaging) insults. This preconditioning effect has been well-documented in skeletal muscle although the role that an increased production of HSPs plays in this adaptation is less well-understood.

Studies in the heart have demonstrated that a short period of hyperthermia or anoxia provides protection of cardiac tissue against damage induced by a subsequent period of normally damaging ischemia and reperfusion.[45–48] This cytoprotection was associated with a sixfold increase in HSP70 content of the preconditioned cardiac tissue. Endurance exercise results in the elevation of HSP70 in the heart by up to 500%,[49] and it has been shown that this cardiac increase in HSP content provides protection against apoptosis, necrosis, and oxidative injury in the cardiac myocyte (see Reference 46 for review). More recently, Marber et al.[50] demonstrated that isolated hearts from transgenic mice, overexpressing HSP70, demonstrated an increased resistance to ischemic injury. This study provided the first direct evidence for a specific role of HSP70 in cytoprotection against muscle damage.

The ability of prior induction of the stress response to protect skeletal muscle against damage has been examined by several workers. The mechanisms of activation of HSP production and subsequent protection against damage in skeletal muscle is summarized in Figure 7.1.

Garramone et al.[51] and Lepore et al.[52] demonstrated that a prior heat stress in rats, which resulted in increased muscle content of HSP70, provided protection to mature skeletal muscle against necrosis induced by ischemia and reperfusion. Muscles from rats that were subjected to a prior heat stress demonstrated a preservation of mitochondrial structure, whereas muscles from nonheated rats demonstrated characteristic changes in mitochondrial structure on analysis by electron microscopy.[51] Studies in cell culture have demonstrated that a prior heat shock provides considerable protection against cell death following hypoxia and re-oxygenation *in vitro*.[53] These workers also demonstrated an enhanced survival of skeletal muscle myoblasts that had been subjected to a prior heat shock, when grafted into the heart.[53] Studies from our laboratory have demonstrated that a prior nondamaging period of exercise results in the rapid production of HSPs in skeletal muscle and that this increased muscle content of HSPs is associated with protection against a subsequent, usually damaging contractile activity in both EDL and soleus muscles.[54] More recent studies have demonstrated a direct role for HSP70 in providing this protection. Preliminary studies indicate that HSP70 provides this protection. Data demonstrated that overexpression of HSP70 in muscles of transgenic mice protects against the secondary loss of force following a period of damaging lengthening contractions and that muscles recovered to pre-exercise maximum force values significantly faster than muscles of wild-type mice.[55]

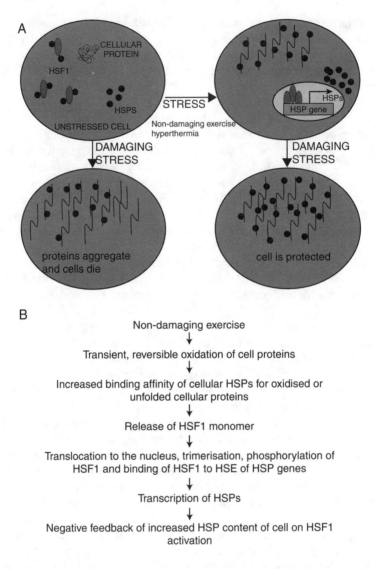

Figure 7.1   Mechanisms of activation of HSF1 and protection against subsequent damage. (From Jackson, M. J., McArdle, A., and McArdle, F., *Proc. Nutr. Soc.*, 57, 301, 1998; Morimoto, R. I., Kroeger, P. E., and Cotto, J. J., in *Stress-Inducible Cellular Responses*, Fiege, U. et al., Eds., Birkhauser Verlag, Basel, 1996, 139–164.)

In contrast, work by Thomas and Noble[56] has demonstrated that an increased muscle content of HSPs has no effect on contractile performance under fatiguing conditions. This is supported by Nosek et al.[57] who have used transgenic mice to demonstrate that an increased muscle content of HSP70 has no effect on the development of muscle fatigue.

VI. The effect of age on skeletal muscle function

Declining muscle strength, muscle wasting, and physical frailty are an accepted part of aging such that by the age of 70, skeletal muscle cross-sectional area is reduced by 25 to 30% and muscle strength is reduced by 30 to 40%.[58] This loss of muscle strength continues to fall by 1 to 2% per year[59] and the decline in power (force multiplied by speed) in the lower leg is 3.5% per year.[59] Such a deficit has a profound impact on the quality of life of older people. Loss of muscle strength leads to instability, a subsequent increased risk of falls, and consequently an increased need for residential care. A large number of healthy older people are at, or near to, functionally important strength-related thresholds and so have lost or are close to losing the ability to carry out everyday tasks.[60] In addition, the age-related decline in muscle bulk and strength contributes to other problems experienced by older people, including an increased susceptibility to hypothermia and increased incidence of incontinence. This loss of muscle function is due to both a loss of muscle fibers and atrophy of the remaining fibers. This atrophy appears to primarily occur in type II (fast, glycolytic) fibers, with the cross-sectional area of type I (slow, oxidative) fibers being relatively well-maintained.[61] Interestingly, type 1 fibers at rest have been shown to contain significantly more HSPs than muscles composed of primarily type 2 fibers.[1] The reduction in the proportion of type II fast muscle fibers results in the muscle moving toward the characteristics of type I or slow muscle fibers. Thus, a slowing of contraction, rate of force development, and thus a reduced ability to accelerate the movement of a limb is observed, amplifying the impact of muscle weakness on stability.[62,63]

In addition, the remaining fibers in aged muscle are weaker than fibers from young individuals. The remaining fibers in muscles of aged mice generate less force per unit cross-sectional area than those of a similar sized muscle in adult mice.[64,65] Thus, loss of muscle fibers accounted for a proportion of the overall force deficit in muscles of aged mice but did not account for all of it, and the force generated per unit cross-sectional area was approximately 25% less than that from muscles of young mice.[66,67] Aged muscle is more susceptible to contraction-induced damage and takes longer to recover from damage. Several studies have demonstrated that skeletal muscle of aged rodents is more susceptible to contraction-induced damage than that of young animals and also takes longer to recover from damage.[64–67] Contraction-induced skeletal muscle damage may occur whenever muscle is exposed to unaccustomed or excessive periods of exercise, but by far the most damaging type of contraction occurs when the muscle is lengthened while activated (also known as eccentric or pliometric contractions). Skeletal muscles of aged rodents are significantly more susceptible to this form of contraction-induced damage.[64,65] In studies from Faulkner's laboratory, the force deficit in muscles of young mice was reduced following a sub-maximal damaging exercise protocol compared with a more severely damaging contraction protocol. Thus, the force deficit in muscles of young mice following the sub-maximal protocol was elevated to a mean of 36%. This sub-maximal protocol highlighted a significantly larger deficit of 57% in muscles of aged mice. This damage also takes considerably longer to repair in old mice such that the muscles from young mice had recovered their ability to generate force by 28 days following the damaging exercise, whereas a significant deficit remained in the muscles of the old mice at 60 days following the exercise.[64,66–68]

Numerous mechanisms have been proposed as theories of aging, but the techniques of molecular genetics and intervention studies in experimental organisms now strongly support a role for free radicals and stress responses in modulation of the processes of aging. The free radical theory of aging was first described by Harman[69] in 1956, and recent successful transgenic[70] and pharmacological interventions[71] to reduce oxidant damage to cells appear to confirm the importance of this mechanism. Stress (or heat shock) proteins are an important part of the cellular responses to free radicals and overexpression of specific heat shock proteins (HSPs) protects cells against free radical-mediated pathologies.[72] Lithgow and colleagues[73,74] have demonstrated that these proteins play an important role in regulation of the aging process in the model organism *Caenorhabditis elegans*. They reported that all naturally occurring mutants that demonstrate increased longevity also showed increased resistance to stress, and analysis of these *Caenorhabditis elegans* mutants indicate that a proportion have an accumulation of HSPs.[73,74]

These data therefore suggest that factors influencing oxidative stress or responses to oxidative stress may play a fundamental role in modulating the aging process and hence be important in the maintenance of muscle function in aging mammalian skeletal muscle.

The ability of cells to induce HSPs following stress is reduced in aged individuals. Tissues for aged animals and blood cells from elderly humans both show a reduced production of stress proteins following thermal stress.[75,76] We have recently demonstrated that this attenuated response occurs in skeletal muscle of aged rats.[77] The production of HSP70 in response to a period of mild, nondamaging contractile activity was severely blunted in comparison with muscles of young animals, although this does not appear to be true in skeletal muscles of rats following a period of whole-body hyperthermia.[78] The ability of muscles of aged rats to adapt following a 10-week training regime on a treadmill seems to be fiber type specific. Work by Naito et al.[79] demonstrated that exercise training of aged rats on a treadmill for 10 weeks results in a similar accumulation of HSP70 in muscles that contain highly oxidative fibers, whereas aging is associated with a blunted expression of HSP70 in predominantly fast skeletal muscles. The lack of adaptation in HSP content in the aged animals may be related to a more general failure of adaptation to stress, particularly in muscles that contain predominantly type 2 fibers. Other data indicate that aged animals have a reduction in their ability to adapt following damaging exercise. Muscles of young rodents show no impairment in muscle force generation following a second period of damaging exercise, whereas muscles from old rodents demonstrate a loss of force similar to that seen following the first period of exercise.[65]

VII. Summary

In summary, data indicate that exercise results in an increased production of HSPs and that an oxidative stress may play a major role in the activation of this response. The increased muscle content of HSPs provides considerable protection to skeletal muscle against damaging stresses. The HSP production in skeletal muscles of aged mammals appears to be attenuated following exercise. This attenuation may play a major role in the development of a functional deficit in muscles of aged mammals.

Acknowledgments

The authors would like to thank *Research into Aging* and the Welcome Trust for financial support of their research.

References

1. Locke, M., Noble, E. G., and Atkinson, B. G., Inducible form of HSP70 is constitutively expressed in a muscle fiber type specific pattern, *Am. J. Physiol.*, 261, C774, 1991.
2. McArdle, A. et al., Contractile activity-induced oxidative stress: cellular origin and adaptive responses, *Am. J. Physiol. Cell. Physiol.*, 280, C621, 2001.
3. Neufer, P. D. et al., Continuous contractile activity induces fiber type specific expression of HSP70 in skeletal muscle, *Am. J. Physiol.*, 271, C1828, 1996.
4. Gething, M. J., Ed., *Guidebook to Molecular Chaperones and Protein-Folding Catalysts*, Oxford University Press, Oxford, 1997.
5. Voellmy, R., Sensing stress and responding to stress, in *Stress-Inducible Cellular Responses*, Feige, U. et al., Eds., Birkhauser, Berlin, 1996, 121–137.
6. Gething, M. J. and Sambrook, J., Protein folding in the cell, *Nature*, 355, 33, 1992.
7. Hoj, P. B., Hoogenraad, N. J., and Hartman, D., Mammalian Cpn60, in *Guidebook to Molecular Chaperones and Protein Folding Catalysts*, Gething, M. J., Ed., Oxford University Press, Oxford, 1997, 197–198.
8. Martinus, R. M. et al., Role of chaperones in the biogenesis and maintenance of mitochondrion, *FASEB J.*, 9, 371, 1995.
9. de Jong, W. W. and Boelens, W. C., Mammalian α-crystallins, in *Guidebook to Molecular Chaperones and Protein Folding Catalysts*, Gething, M. J., Ed., Oxford University Press, Oxford, 1997, 288–290.
10. Bennardini, F., Wrzosek, A., and Chiesi, M., αB-crystallin in cardiac tissue. Association with actin and desmin filaments, *Circ. Res.*, 71, 288, 1992.
11. Klemenz, R. et al., Alpha-B-crystallin in a small heat shock protein, *Proc. Natl. Acad. Sci. U.S.A.*, 88, 3652, 1991.
12. Das, D. K. et al., Signal transduction pathway leading to HSP27 and HSP70 gene expression during myocardial adaptation to stress, in *Stress of Life from Molecules to Man*, Csermely, P., Ed., *Annals of the New York Academy*, 851, 129–138, 1998.
13. Goldspink, G., Cellular and molecular aspects of adaptation in skeletal muscle, in *Strength and Power in Sport*, Komi, P. V., Ed., Blackwell Science, 1994, 211–229.
14. Locke, M., Noble, E. G., and Atkinson, B. G., Exercising mammals synthesise stress proteins, *Am. J. Physiol.*, 258, C723, 1990.

15. Salo, D. C., Donovan, C. M., and Davies, K. J., HSP70 and other possible heat shock or oxidative stress proteins are induced in skeletal muscle, heart and liver during exercise, *Free Rad. Biol. Med.*, 11, 239, 1991.

16. Neufer, P. D., Ordway, G. A., and Williams, R. S., Transient regulation of c-fos, alpha B-crystallin, and HSP70 in muscle during recovery from contractile activity, *Am. J. Physiol.*, 274, C341, 1998.

17. Ornatsky, O. I., Connor, M. K., and Hood, D. A., Expression of stress proteins and mitochondrial chaperonins in chronically stimulated skeletal muscle, *Biochem. J.*, 311, 119, 1995.

18. Ecochard, L. et al., Skeletal muscle HSP72 level during endurance training: influence of peripheral arterial insufficiency, *Pflugers Arch.*, 440, 918, 2000.

19. Gonzalez, B., Hernando, R., and Manso, R., Stress proteins of 70 kDa in chronically exercised skeletal muscle, *Pflugers Arch.*, 440, 42, 2000.

20. Samelman, T. R., Shiry, L. J., and Cameron, D. F., Endurance training increases the expression of mitochondrial and nuclear encoded cytochrome c oxidase subunits and heat shock proteins in rat skeletal muscle, *Eur. J. Appl. Physiol.*, 83, 22, 2000.

21. Puntschart, A. et al., HSP70 expression in human skeletal muscle after exercise, *Acta Physiol. Scand.*, 157, 411, 1996.

22. Febbraio, M. A. and Koukoulas, I., HSP72 gene expression progressively increases in human skeletal muscle during prolonged, exhaustive exercise, *J. Appl. Physiol.*, 89, 1055, 2000.

23. Khassaf, M. et al., Time course of responses of human skeletal muscle to oxidative stress induced by nondamaging exercise, *J. Appl. Physiol.*, 90, 1031, 2001.

24. Liu, Y. et al., Human skeletal muscle HSP70 response to training in highly trained rowers, *J. Appl. Physiol.*, 86, 101, 1999.

25. Brooks, G. A. et al., Tissue temperature and whole-animal oxygen consumption after exercise, *Am. J. Physiol.*, 221, 427, 1971.

26. Skidmore, R. et al., HSP70 induction during exercise and heat stress in rats: role of internal temperature, *Am. J. Physiol.*, 268, R92, 1995.

27. Boveris, A. and Chance, B., The mitochondrial generation of hydrogen peroxide. General properties and effect of hyperbaric oxygen, *Biochem. J.*, 134, 707, 1973.

28. Cadenas, E. et al., Production of superoxide radicals and hydrogen peroxide by NADH-ubiquinone reductase and ubiquinol-cytochrome c reductase from beef-heart mitochondria, *Arch. Biochem. Biophys.*, 180, 248, 1977.

29. McArdle, A. and Jackson, M. J., Exercise, oxidative stress and aging, *J. Anat.*, 197, 539, 2000.

30. Shigenaga, M. K., Hagen, T. M., and Ames, B. N., Oxidative damage and mitochondrial decay in aging, *Proc. Natl. Acad. Sci., U.S.A.*, 91, 10771, 1994.

31. Reid, M. B. et al., Reactive oxygen in skeletal muscle. II. Extracellular release of free radicals, *J. Appl. Physiol.*, 73, 1805, 1992.

32. O'Neill, C. A. et al., Production of hydroxyl radicals in contracting skeletal muscle of cats, *J. Appl. Physiol.*, 81, 1197, 1996.

33. Balon, T. W. and Nadler, J. L., Nitric oxide release is present from incubated skeletal muscle preparations, *J. Appl. Physiol.*, 77, 2519, 1994.

34. Ohno, H. et al., Superoxide dismutases in exercise and disease, in *Exercise and Oxygen Toxicity*, Sen, C. K., Packer, L., and Hanninen, O., Eds., Elsevier, Amsterdam, 1994, 127–162.

35. Alessio, H. M. and Goldfarb, A. H., Lipid peroxidation and scavenger enzymes during exercise: adaptive response to training, *J. Appl. Physiol.*, 64, 1333, 1988.

36. Caldarera, C. M., Guanieri, C., and Lazzari, F., Catalase and peroxidase activity of cardiac muscle, *Bull. Soc. Ital. Biol. Sper.*, 49, 72, 1973.

37. Vincent, H. K. et al., Short-term exercise training improves diaphragm antioxidant capacity and endurance, *Eur. J. Appl. Physiol.*, 81, 67, 2000.

38. Jackson, M. J., McArdle, A., and McArdle, F., Antioxidant micronutrients and gene expression, *Proc. Nutr. Soc.*, 57, 301, 1998.

39. Jackson, M. J., Edwards, R. H. T., and Symons, M. C. R., Electron spin resonance studies of intact mammalian skeletal muscle, *Biochim. Biophys. Acta*, 847, 185, 1985.

40. Freeman, M. L. et al., Characterisation of a signal generated by oxidation of protein thiols that activates the heat shock transcription factor, *J. Cell. Physiol.*, 164, 356, 1995.

41. Khassaf, M. et al., Effects of vitamin E or β-carotene on response of human muscle to oxidative stress, *Free Rad. Biol. Med.*, 27, S37, 1999.

42. Jackson, M. J. et al., Vitamin C supplements suppress the stress response in human muscle, *Free Rad. Biol. Med.*, 27, S36, 1999.

43. Morimoto, R. I., Kroeger, P. E., and Cotto, J. J., The transcriptional regulation of heat shock genes: a plethora of heat shock factors and regulatory conditions, in *Stress-Inducible Cellular Responses*, Fiege, U. et al., Eds., Birkhauser Verlag, Basel, 1996, 139–164.

44. McArdle, A. and Jackson, M. J., Intracellular mechanisms involved in skeletal muscle damage, in *Muscle Damage*, Salmons, S., Ed., Oxford University Press, Oxford, 1997, 90–106.

45. Yellon, D. M. et al., The protective role of heat stress in the ischaemic and reperfused rabbit myocardium, *Mol. Cell. Cardiol.*, 24, 895, 1992.

46. Marber, M. S., Stress proteins and myocardial protection, *Clin. Sci.*, 86, 375, 1994.

47. Currie, R. W. et al., Heat-shock response is associated with enhanced postischaemic ventricular recovery, *Circ. Res.*, 63, 543, 1988.

48. Donnely, T. J. et al., Heat shock protein induction in rat hearts. A role for improved myocardial salvage after ischaemia and reperfusion?, *Circulation*, 85, 769, 1992.

49. Powers, S. K., Locke, M., and Demirel, H. A., Exercise, heat shock proteins and myocardial protection, *Med. Sci. Sports Exerc.*, 33, 386, 2001.

50. Marber, M. S. et al., Overexpression of the rat inducible 70-kD heat stress protein in a transgenic mouse increases the resistance of the heart to ischemic injury, *J. Clin. Invest.*, 95, 1446, 1995.

51. Garramone, R. R. et al., Reduction of skeletal muscle injury through stress conditioning using the heat shock response, *Plast. Reconstr. Surg.*, 93, 1242, 1994.

52. Lepore, D. A. et al., Prior heat stress improves survival of ischaemic-reperfused skeletal muscle *in vivo*, *Muscle Nerve*, 23, 1847, 2000.

53. Suzuki, K. et al., Heat shock treatment enhances graft cell survival in skeletal muscle myoblast transplantation to the heart, *Circulation*, 102 (Suppl. 3), III216, 2000.

54. McArdle, A., McArdle, C., and Jackson, M. J., Stress proteins and protection of skeletal muscle against contraction-induced skeletal muscle damage in anaesthetised mice, *J. Physiol.*, 499P, 9P 1997.

55. McArdle, A. et al., Increased expression of HSP70 protects skeletal muscle against damage following lengthening contractions, *FASEB J.*, 14(4), A50, 76.5, 2000.

56. Thomas, J. A. and Noble, E. G., Heat shock does not attenuate low-frequency fatigue, *Can. J. Physiol. Pharmacol.*, 77, 64, 1999.

57. Nosek, T. M. et al., Functional properties of skeletal muscle from transgenic animals with upregulated heat shock protein 70, *Physiol. Genomics*, 4, 25, 2000.

58. Porter, M. M., Vandervoort, A. A., and Lexell, J., Aging of human muscle: structure, function and adaptability, *Scand. J. Med. Sci. Sports*, 5, 129, 1995.

59. Skelton, D. A. et al., Strength, power and related functional ability of healthy people aged 65–89 years, *Age Ageing*, 23, 371, 1994.
60. Young, A. and Skelton, D. A., Applied physiology of strength and power in old age, *Int. J. Sports Med.*, 15, 149, 1994.
61. Lexell, J., Taylor, C. C., and Sjostrom, M., What is the cause of the aging atrophy? Total number, size and proportion of different fiber types studied in whole vastus lateralis muscle from 15- to 83-year-old men, *J. Neurol. Sci.*, 84, 275, 1988.
62. Larson, L., Grimby, G., and Karlsson, J., Muscle strength and speed of movement in relation to age and muscle morphology, *J. Appl. Physiol.*, 46, 451, 1979.
63. Stanley, S. N. and Taylor, N. A. S., Isokinematic muscle mechanics in four groups of women of increasing age, *Eur. J. Appl. Physiol.*, 66, 198, 1993.
64. Brooks, S. V. and Faulkner, J. A., Contractile properties of skeletal muscles from young, adult and aged mice, *J. Physiol.*, 404, 71, 1988.
65. McBride, T. A., Gorin, F. A., and Carlsen, R. C., Prolonged recovery and reduced adaptation in aged rat muscle following eccentric exercise, *Mech. Ageing Dev.*, 83, 185, 1995.
66. Faulkner, J. A., Brooks, S. V., and Zerba, E., Skeletal muscle weakness and fatigue in old age: underlying mechanisms, *Annu. Rev. Gerontol. Geriatr.*, 10, 147, 1990.
67. Faulkner, J. A., Brooks, S. V., and Zerba, E., Skeletal muscle weakness, fatigue, and injury: inevitable concomitants of aging?, *Hermes (Leuven)*, XXI (2-3), 269, 1990.
68. Zerba, E., Komorowski, T. E., and Faulkner, J. A., Free radical injury to skeletal muscles of young, adult, and old mice, *Am. J. Physiol.*, 258(3 Pt. 1), C429, 1990.
69. Harman, D., Ageing: theory based on free radical and radiation chemistry, *J. Gerontol.*, 11, 298, 1956.
70. Orr, W. C. and Sohal, R. S., Extension of life-span by overexpression of superoxide dismutase and catalase in drosophila melanogaster, *Science*, 263 (5150), 1128, 1994.
71. Melov, S. et al., Extension of life-span with superoxide dismutase/catalase mimetics, *Science*, 289(5484), 1567, 2000.
72. Marber, M. S. et al., Overexpression of the rat inducible 70-kD heat stress protein in a transgenic mouse increases the resistance of the heart to ischemic injury, *J. Clin. Invest.*, 95(4), 1446, 1995.
73. Walker, G. A., Walker, D. W., and Lithgow, G. J., Genes that determine both thermotolerance and rate of aging in *Caenorhabditis elegans*, *Ann. N.Y. Acad. Sci.*, 851, 444, 1998.
74. Sampayo, J. N., Jenkins, N. L., and Lithgow, G. J., Using stress resistance to isolate novel longevity mutations in *Caenorhabditis elegans*, *Ann. N.Y. Acad. Sci.*, 908, 324, 2000.
75. Liu, A. Y. C. et al., Attenuated heat shock transcriptional response in ageing: molecular mechanism and implication in the biology of ageing, in *Stress-Inducible Cellular Responses*, Feige, U. et al., Eds., Birkhauser, Berlin, 1996, 393–408.
76. Rao, D. V., Watson. K., and Jones, G. L., Age related attenuation in the expression of the major heat shock proteins in human peripheral lymphocytes, *Mech. Aging Dev.*, 107, 105, 1999.
77. Vasilaki, A., Jackson, M. J., and McArdle, A., Attenuated HSP70 response in skeletal muscle of aged rats following contractile activity, *Muscle and Nerve* (submitted for publication).
78. Locke, M., Heat shock transcription factor activation and HSP72 accumulation in aged skeletal muscle, *Cell Stress Chaperones*, 5, 45, 2000.
79. Naito, H. et al., Exercise training increases heat shock protein in skeletal muscles of old rats, *Med. Sci. Sports Exerc.*, 33, 729, 2001.

chapter eight

Stress proteins and the mitochondrion

David A. Hood, Arne A. Rungi, Marco Colavecchia,
Joesph W. Gordon, and Jeremy J. Schneider

Contents

I. Introduction

Stress proteins are intimately involved in a wide variety of cellular events, including organelle biogenesis. Recent work has indicated that the expression of a number of these can be induced by the stress of exercise, and that this is fundamental for the formation of new mitochondria. The synthesis of mitochondria (i.e., mitochondrial biogenesis) is a hallmark adaptation that is a well-established response to regular physical activity in skeletal muscle. In addition, mitochondrial biogenesis also occurs in cardiac muscle in the presence of high levels of thyroid hormone. Failure of the normal sequence of events that occur during mitochondrial biogenesis can lead to mitochondrial disease, often most visible in tissues requiring a high ATP supply (e.g., brain, heart, muscle). This chapter provides an overview of mitochondrial biogenesis and focuses on the essential roles of stress proteins in this process, particularly in muscle and heart, when subjected to stimuli that alter mitochondrial content.

II. Overview of mitochondrial biogenesis

Mitochondria are highly specialized organelles that play a vital role in generating the ATP required for life-sustaining cellular activities. Their structure consists of two lipid bilayers, as well as an internal membrane system and a matrix. More than 100 proteins embedded within the membranes, or localized in the matrix, are responsible for ATP synthesis. A variety of physiological conditions exist[1] in which biogenesis is stimulated, resulting in an elevated mitochondrial content, in addition to changes in organelle composition. This process is complex because it requires the cooperation of the nuclear and the mitochondrial genomes. The mitochondrion is unique because it possesses its own DNA and protein synthetic apparatus, which are distinct from those found in the nucleus and cytoplasm, respectively. As reviewed recently,[2] mitochondrial biogenesis involves at least five events: (1) the synthesis of phospholipids to expand the membrane size; (2) the replication and transcription of mitochondrial DNA, and mRNA translation into protein; (3) the transcription and translation of nuclear genes; (4) the chaperoning and subsequent import of nuclear gene products into the newly forming organelle; and (5) the refolding and correct assembly of proteins and lipids into a functional stoichiometry. Stress proteins are heavily implicated in steps 4 and 5 (above) and are located in both the cytosol and within the mitochondrial matrix.

III. Cytosolic chaperones and protein targeting to mitochondria

The synthesis of nuclear-encoded mitochondrial proteins largely takes place on free ribosomes in the cytosol. These are known as "precursor" proteins because they are often fabricated with an N-terminal extension that is removed upon entry into the organelle. Once translated, the precursor protein is released and directed to the mitochondria by molecular chaperones. The most important of these are believed to be the cytosolic heat shock protein 70 kDa (HSP70) and mitochondrial import stimulation factor (MSF; Figure 8.1).[3] The ability of these molecules to interact with the precursor stems from the presence of a specific amino acid sequence within the precursor polypeptide. This sequence is composed of positively charged amino acids and contains the necessary information to target the protein to the mitochondria. It may be located within the N-terminal signal region or within the sequence of the mature protein itself. As is the case with other cellular proteins, mitochondrial

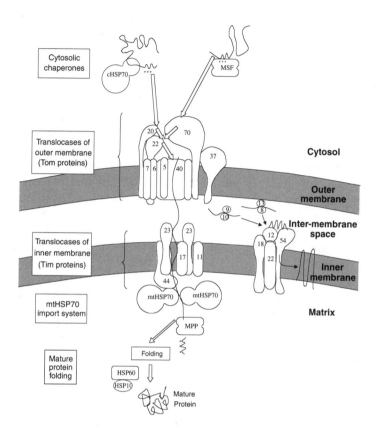

Figure 8.1 Protein import into mitochondria. Precursor proteins made in the cytosol are chaperoned to the import machinery by a cytosolic chaperone, either cytosolic HSP70 (cHSP70) or mitochondrial import stimulating factor (MSF). The precursor interacts with either the Tom20-22 or Tom70-37 heterodimers, and is then directed to the general import pore consisting of the main outer membrane protein channel Tom40. Precursors destined for the matrix interact with Tim 17-23, which form a channel in the inner membrane. The matrix chaperone mtHSP70, which is anchored to the Tim complex by Tim44, pulls in the precursor in an ATP-dependent fashion. The signal sequence is cleaved by the mitochondrial processing peptidase (MPP). Subsequently, the mature protein is refolded by matrix chaperonins HSP60 and cpn10. Muscle contractile activity (i.e., exercise) leads to an increased expression of Tom20, cHSP70, MSF, HSP60, cpn10, and mtHSP70, likely facilitating the import of proteins into the organelle during mitochondrial biogenesis. Proteins destined for the inner membrane require the use of another set of proteins, initially involving Tim9, Tim10, Tim13, and Tim8. Nothing is currently known about this latter pathway in mammalian cells. The number within each import machinery component refers to its size in kilodaltons (kDa). See the text for details. (Adapted from Pfanner, N. and Meijer, M., *Curr. Biol.*, 7, R100, 1997; Koehler, C. M., *FEBS Lett.*, 476, 27, 2000.)

precursors cannot undergo translocation into the mitochondria unless they are maintained in an unfolded state. The task of destabilizing the structure of precursors before they are imported is also carried out by molecular chaperones. Both HSP70 and MSF exploit the inherent instability of the precursor so that it can properly interact with the mitochondrial protein import machinery (see below) and be less susceptible to aggregating prior to translocation.

Despite their overall functional similarities, these molecular chaperones possess certain unique characteristics. MSF, a heterodimer consisting of 30-kDa and 32-kDa subunits, specifically recognizes mitochondrial targeted proteins and functions to unfold and bind these polypeptides.[6] The presence of basic amino acids in the precursor protein is critical for MSF binding.[7] In addition to recognizing matrix targeting N-terminal presequences, MSF also plays a role in the import of mitochondrial outer membrane proteins containing hydrophobic internal targeting sequences.[7] MSF directs precursor proteins predominantly to Tom70-Tom37, forming a stable complex. ATP hydrolysis by MSF releases it, leaving the precursor bound to the outer membrane import machinery.[8] In contrast, HSP70 directs precursors to the import machinery in the absence of a cellular energy requirement.[9] HSP70 generally binds unfolded proteins in the cytosol and thus stabilizes a variety of proteins, not only those destined for the mitochondrion. Unlike MSF, HSP70 does not directly interact with the import machinery, and import occurs independent of the Tom70-Tom37 receptor subcomplex. Instead, precursors bound to HSP70 are transferred to the Tom20-Tom22 subcomplex in an ATP-independent fashion.[3] The binding of precursor to Tom20-Tom22 may involve an interaction of the positively charged presequence of the precursor to negatively charged portions of the Tom20-Tom22 complex. Thus, HSP70 and MSF present precursors to separate subcomplexes of the mitochondrial outer membrane import machinery.

IV. The import machinery of the outer membrane

The proteins comprising the multimeric receptor complex in the outer mitochondrial membrane have been termed translocases of the outer membrane (Tom; Figure 8.1). Following delivery of the precursor to the Tom machinery by molecular chaperones, the receptor complexes assist in the movement of the precursor into the organelle. As noted above, the receptors Tom20 and Tom22 appear to be the primary targets for HSP70-taxied precursors,[3] whereas proteins interacting with MSF are largely directed to the Tom70-Tom37 heterodimer.[10] The precursor is then directed to the multi-subunit general insertion pore (GIP) consisting of a 400-kDa complex comprising the protein channel Tom40,[11,12] and including the smaller proteins Tom5, Tom6, and Tom7. Tom5 mediates the transfer of the precursor from Tom20/Tom22 to Tom40, while Tom6 and Tom7 stabilize and destabilize, respectively, the interactions between Tom22 and Tom40.[13] Movement of the precursor from the outer to the inner membrane appears to rely on a series of interactions of increasing affinity between the positively charged presequence and acidic patches on the surface of Tom and Tim (translocases of the inner membrane) proteins, termed the "acid chain hypothesis."[14]

As with other components of the import machinery, including the stress proteins involved in chaperoning functions, the majority of work examining the Tom proteins has been performed in *Saccharomyces cerevisiae* and *Neurospora crassa*.[2] However, the kinetics of protein import into mammalian mitochondria are now becoming more clearly defined. For example, it was recently shown that Tom20 plays an important role in the import of matrix-destined precursors in C2C12 skeletal muscle cell mitochondria.[15] Increases in the expression of Tom20 in these cells brought about by either thyroid hormone treatment or transfection with a mammalian expression construct resulted in comparable increases in import of the citric acid cycle enzyme malate dehydrogenase (MDH). Conversely, inhibition of Tom20 expression using antisense oligonucleotides, or inhibition of function using a Tom20 antibody, led to equivalent decreases in MDH import.

V. The import machinery of the inner membrane

As the preprotein traverses the outer membrane and intermembrane space, it is continually maintained in an unfolded, import-competent conformation.[16] Entry of the positively charged presequence into the inner membrane channel is dependent on the electrophoretic effect of the membrane potential ($\Delta\psi \approx -120$ mV inside), as well as interaction with anionic phospholipids (i.e., cardiolipin) in the inner membrane which stabilizes the presequence and directs it[17] to the inner membrane protein complexes. These are termed translocases of the inner membrane (Tim) proteins. Unlike the Tom proteins, Tim proteins are present in lesser quantities. It is estimated that Tom proteins are in twofold molar excess over their inner membrane counterparts.[18] The transferring of precursors from the Tom to the Tim proteins is facilitated by the fact that both complexes are located in areas of close contact between the inner and outer membranes.[19] Two of the Tim proteins (Tim23 and Tim17) act as integral membrane proteins. In a similar fashion to the Tom proteins, Tim17 and Tim23 are capable of binding precursor molecules and directing them along the import pathway. It is believed that Tim23 and Tim17 form an aqueous channel through which precursors can pass on their way to the matrix.

Although Tim17 and Tim23 play an instrumental role in the import of matrix-destined preproteins, neither appears to contribute to the targeting of carrier proteins residing in the inner membrane. Proteins found within the inner membrane have either N-terminal or internal mitochondrial targeting sequences, as well as adjacent topogenic signals that are hydrophobic cores flanked by hydrophilic amino acids. These signals ensure sorting of the precursor to the inner membrane and determine its orientation.[20] Proteins are transferred from the Tom complex, through the intermembrane space, directly to the inner membrane lipid bilayer. A new set of protein import machinery components identified in yeast (Tim8, Tim9, Tim10, Tim12, Tim13) directs precursor proteins across the aqueous intermembrane space to interact with a separate inner membrane import machinery complex consisting of Tim18, Tim22, and Tim54 (Figure 8.1).[21] Virtually nothing is known about the functions of these proteins in mammalian cells, except that the loss of the human Tim8 homologue results in the Mohr-Tranebjaerg syndrome. This condition is characterized by dystonia, muscle weakness, deafness, and blindness,[22] probably as a result of a reduced abundance of some inner membrane proteins that are critical for the functions sensory and muscular systems.[5]

VI. Translocation from the inner membrane to the matrix

Precursors destined for the matrix side are bound to the stress protein termed mitochondrial HSP70 (mtHSP70). mtHSP70 uses Tim44 as an anchor to "pull" the precursor across the inner membrane in an ATP-dependent fashion,[5] and it has been suggested that mtHSP70 plays a regulatory role in protein import.[23] Disruption of the interaction between Tim44 and mtHSP70 impairs protein import[24,25] and delays the folding of those proteins that are imported.[26] In addition to ATP, one other factor has been identified as being necessary for efficient import. Mge1 (mitochondrial GRPE) acts together with Tim44 and mtHSP70.[27] Mge1 mediates the association of nucleotides with mtHSP70, which is necessary for mtHSP70 function.[28] There have been two hypotheses put forth to explain the function of mtHSP70 in "pulling" the precursor protein into the matrix.[29] The first—the "translocation motor" model—suggests that mtHSP70 undergoes a conformational change to pull the preprotein into the matrix, and then binds further along the protein prior to exerting a subsequent pulling action. This model is supported by data showing that mtHSP70 promotes unfolding of the remaining portion of the protein being imported.[30] This "unfoldase" function suggests that mtHSP70 pulls on the protein, something that could only be accomplished if mtHSP70 underwent a conformational change while anchored to the inner membrane.

In the "Brownian ratchet" model, mtHSP70 binds the presequence and then dissociates from the inner membrane. It is postulated that the binding of mtHSP70 functions to prevent the backward movement of the precursor out of the mitochondria. In time, the precursor diffuses in the only direction that it is permitted to—into the matrix. In this model, binding of other mtHSP70 molecules along the length of the protein may be involved. Evidence for this model is provided by experiments that show that mtHSP70 prevents the backsliding of proteins through translocation channels.[31] It should be noted that it is possible that mtHSP70 acts in a manner that satisfies both models.[32] In other words, an mtHSP70 molecule may initially bind to the presequence, fulfilling the trapping role, and then a second mtHSP70 molecule could bind further along the protein, acting as a motor to pull in the protein.

VII. Precursor processing in the matrix

As the precursor enters the matrix, the presequence is cleaved by the mito-chondrial processing peptidase (MPP; Figure 8.1).[33,34] Such cleavage results in the mature protein possessing a lower molecular weight than its nonim-ported counterpart. MPP activity is dependent on divalent metal ions such as Mn^{2+}, and it is enhanced by processing enhancing protein (PEP).[35] The mature protein must then be reconstituted into its three-dimensional struc-ture to ensure its correct function. mtHSP70 has been implicated as a key participant in stabilizing and folding many matrix proteins,[36] including those encoded by the mitochondrial genome.[37] As a chaperone, mtHSP70 functions together in a complex with Mge1 and Mdj1.[38] Assisting mtHSP70 in its chaperone function are HSP60 and chaperonin 10 (cpn10), which serve to further fold some of the proteins already partially folded by mtHSP70.[39] Indeed, the majority of mitochondrial proteins require the assistance of HSP60 to attain their native conformation.[39,40] The folding of mitochondrial precursor proteins is the third and final ATP-dependent step in the import process. It is believed that partially folded precursors enter the cylindrical cavity of HSP60, bind to its wall, and through various cycles of clasp and release, are allowed to take on their mature conformation. HSP60 releases the protein into the matrix with the aid of the co-chaperone cpn10.[41] Thus, HSP60 promotes the folding, assembly, and stabilization of mitochondrial proteins, including those comprising the electron transport chain. This stress protein is so important that the pathological features of some forms of human mitochondrial disease have been linked to reductions in steady-state levels of HSP60.[42,43] The important role played by this chaperone in protein folding is also underscored by the finding that it is inducible during conditions of altered mitochondrial biogenesis[44] and during the imposition of a mitochondrial stress.[45]

VIII. Stress proteins and mitochondrial biogenesis

Contractile activity-induced cell stress is a powerful stimulus for mitochon-drial biogenesis. This is mediated by the transduction of a putative physio-logical signal to the nucleus, resulting in enhanced expression of nuclear genes encoding transcription factors, stress proteins, and those proteins directly involved in mitochondrial respiration.[2] Interestingly, the alteration in mitochondrial phenotype brought about by contractile activity is coinci-dent with changes in the expression of import machinery components. For example, mitochondrial cardiolipin content increases in skeletal muscle between 5 and 10 days of chronic stimulation.[46] Chronic contractile activity ranging from 5 to 14 days elicits increases in the protein levels of Tom20, HSP60, mtHSP70, and the large subunit of MSF.[44,47,48] Importantly, these changes were accompanied by accelerated rates of precursor import into the mitochondrial matrix, including the import of mitochondrial transcription factor A (Tfam).[48] This transcription factor is responsible for the replication and transcription of mitochondrial DNA, a process that is vital for the synthesis of respiratory chain subunits. Thus, these fundamental adaptations in the mitochondrial protein import pathway not only permit an expansion of the mitochondrial reticulum in striated muscle cells, but they also allow for an overall improvement in mitochondrial energy production.

The administration of thyroid hormone (T_3), both *in vivo* and *in vitro*, has also been shown to influence organelle biogenesis, particularly in the heart. Modifications in the expression of mitochondrial import machinery components (e.g., Tom20) and stress proteins (e.g., mtHSP70 and HSP60) are key adaptations that occur in response to T_3.[49,50] These changes are also coincident with an enhanced rate of matrix protein import in cardiac mitochondria. The T_3-induced increase in import rate and mtHSP70 content occurred in the absence of any changes in MPP activity,[50] suggesting that this molecule does not play a role in determining the overall kinetics of protein import into mitochondria.

IX. Stress proteins and mitochondrial disease

While many studies have investigated the expression of stress proteins in yeast with oxidative phosphorylation defects, few have examined this expression in human mitochondrial disease. In a case study of a patient who exhibited multiple mitochondrial enzymatic defects, the two stress proteins investigated were differentially expressed.[42] The amount of matrix-localized HSP60 was reduced by 80%, whereas cytosolic HSP70 was no different than the control. This difference was attributed to decreased synthesis and inefficient mitochondrial import of HSP60 in the patient.[43] HSP60-deficient fibroblasts showed an altered mitochondrial morphology not found in patients with mitochondrial encephalopathy with lactic acid syndrome (MELAS) or COX deficiency. The partial deficiency of HSP60 was thought to be the primary cause of the disease for two main reasons: (1) the deficient enzymes were located on the inner membrane or in the matrix, the assembly of which relies on HSP60, whereas the activity of outer membrane enzymes was normal; and (2) family history suggested an autosomal recessive mode of inheritance, indicating that the primary defects were nuclear in origin. Immunolabeling of HSP60 in muscle fibers of patients with mitochondrial encephalomyopathies revealed differential expression of the protein in the subsarcolemmal (SS) and intermyofibrillar (IMF) mitochondria.[51] In fact, the SS mitochondrial subfraction contained negligible amounts of HSP60, whereas the IMF mitochondria appeared to overexpress the protein. The lack of HSP60 in ragged red fibers suggests that this chaperone protein is implicated in the normal biogenesis of mitochondria in muscle. Recently, we observed a modest decrease in the synthesis but no change in the rate of mitochondrial import of HSP60 in the skin fibroblasts of a patient with multiple mitochondrial disease of nuclear origin.[52] Thus, these studies indicate that stress proteins may, in some but not all cases, have a significant role in the pathogenesis of mitochondrial disease. More investigation is required not only on HSP60, but also the other mitochondrial stress proteins mtHSP70 and cpn10.

In recent years, mitochondria with depleted mtDNA have been used to provide insight into some of the molecular events involved in the pathogenesis of mitochondrial disease. To achieve a state of depleted mtDNA (rho⁻), cells are grown in the presence of ethidium bromide. Continued treatment may result in cells devoid of mtDNA, in which case they are termed rho^0 cells.[53] When the level of mtDNA in a cell is reduced, less mtRNA will be transcribed and, consequently, there will be a decrease in the expression of mitochondrially encoded proteins. This, in turn, leads to dysfunction of the respiratory complexes and a decrease in cellular oxidative capacity.[54] As a compensatory response to this state of metabolic stress, many nuclear genes involved in oxidative phosphorylation, including cytochrome c, cytochrome c oxidase subunit IV, and the adenine nucleotide translocase are up-regulated.[55] In this case, the mRNA encoding the cytosolic stress protein HSP70 remained unaffected. In addition, protein levels of HSP70 mirrored the expression at the mRNA levels.[45] This effect was also observed with mtHSP70. In contrast, the other matrix chaperones HSP60 and cpn10 were induced in the rho^0 cells, indicating that distinct regulatory mechanisms regulate the expression of the various cytosolic and mitochondrial stress proteins.[45] This is not surprising given that they are products of different genes and are influenced differentially by various cellular stressors.

X. Summary and conclusions

Stress proteins perform vital functions in maintaining cell viability. This is in part related to their important roles in the biogenesis of mitochondria. The most important of these identified so far include cytosolic HSP70, MSF, mtHSP70, HSP60, and cpn10. Cytosolic HSP70 is the best characterized of these, both functionally and biochemically. These proteins are induced independently by both exercise and thyroid hormone and serve to maintain mitochondrial integrity by enhancing protein translocation from the cytosol to the appropriate mitochondrial compartment. Areas for future work include investigations on (1) the regulation of the expression of these genes at transcriptional and post-transcriptional levels under a variety of physiological conditions and in mitochondrial disease; and (2) the functions of the intramitochondrial stress proteins in protein refolding, degradation, and export of newly synthesized mtDNA gene products to the inner membrane.

References

1. Hood, D. A. et al., Mitochondrial biogenesis in striated muscle, *Can. J. Appl. Physiol.*, 19, 12, 1994.
2. Hood, D. A., Invited review: Contractile activity-induced mitochondrial biogenesis in skeletal muscle, *J. Appl. Physiol.*, 90, 1137, 2001.

3. Mihara, K. and Omura, T., Cytoplasmic chaperones in precursor targeting to mitochondria: the role of MSF and HSP70, *Trends Cell Biol.*, 6, 104, 1996.
4. Pfanner, N. and Meijer, M., Mitochondrial biogenesis: the Tom and Tim machine, *Curr. Biol.*, 7, R100, 1997.
5. Koehler, C. M., Protein translocation pathways of the mitochondrion, *FEBS Lett.*, 476, 27, 2000.
6. Hachiya, N. et al., A mitochondrial import factor purified from rat liver cytosol is an ATP-dependent conformational modulator for precursor proteins, *EMBO J.*, 12, 1579, 1993.
7. Komiya, T. et al., Recognition of mitochondria-targeting signals by a cytosolic import stimulation factor, MSF, *J. Biol. Chem.*, 269, 30893, 1994.
8. Komiya, T. et al., Binding of mitochondrial precursor proteins to the cytoplasmic domains of the import receptors Tom70 and Tom20 is determined by cytoplasmic chaperones, *EMBO J.*, 16, 4267, 1997.
9. Komiya, T., Sakaguchi, M., and Mihara, K., Cytoplasmic chaperones determine the targeting pathway of precursor proteins to mitochondria, *EMBO J.*, 15, 399, 1996.
10. Hachiya, N. et al., Reconstitution of the initial steps of mitochondrial protein import, *Nature,* 376, 705, 1995.
11. Hill, K. et al., Tom40 forms the hydrophilic channel of the mitochondrial import pore for preproteins, *Nature,* 395, 516, 1998.
12. Kunkele, K. P. et al., The isolated complex of the translocase of the outer membrane of mitochondria. Characterization of the cation-selective and voltage-gated preprotein-conducting pore, *J. Biol. Chem.*, 273, 31032, 1998.
13. Voos, W. et al., Mechanisms of protein translocation into mitochondria, *Biochim. Biophys. Acta,* 1422, 235, 1999.
14. Schatz, G., Just follow the acid chain, *Nature,* 388, 121, 1997.
15. Grey J. Y. et al., Tom20-mediated mitochondrial protein import in muscle cells during differentiation, *Am. J. Physiol.*, 279, C1393, 2000.
16. Rassow, J. et al., Polypeptides traverse the mitochondrial envelope in an extended state, *FEBS Lett.*, 275, 190, 1990.
17. Leenhouts, J. M. et al., The N-terminal half of a mitochondrial presequence peptide inserts into cardiolipin-containing membranes. Consequences for the action of a transmembrane potential, *FEBS Lett.*, 388, 34, 1996.
18. Sirrenberg, C. et al., Functional cooperation and stoichiometry of protein translocases of the outer and inner membranes of mitochondria, *J. Biol. Chem.*, 272, 29963, 1997.
19. Pfanner, N. et al., Contact sites between inner and outer membranes: structure and role in protein translocation into the mitochondria, *Biochim. Biophys. Acta,* 1018, 239, 1990.
20. Stuart, R. A. and Neupert, W., Topogenesis of inner membrane proteins of mitochondria, *Trends Biochem. Sci.*, 21, 261, 1996.
21. Pfanner, N., Mitochondrial import: crossing the aqueous intermembrane space, *Curr. Biol.*, 8, R262, 1998.
22. Koehler, C. M. et al., Human deafness dystonia syndrome is a mitochondrial disease, *Proc. Natl. Acad. Sci. U.S.A.*, 96, 2141, 1999.
23. Dekker, P. J. et al., The Tim core complex defines the number of mitochondrial translocation contact sites and can hold arrested preproteins in the absence of matrix HSP70-Tim44, *EMBO J.*, 16, 5408, 1997.

24. Bömer, U. et al., Separation of structural and dynamic functions of the mitochondrial translocase: Tim44 is crucial for the inner membrane import sites in translocation of tightly folded domains, but not of loosely folded preproteins, *EMBO J.*, 17, 4226, 1998.

25. Merlin, A. et al., The J-related segment of Tim44 is essential for cell viability: a mutant Tim44 remains in the mitochondrial import site, but inefficiently recruits mtHSP70 and impairs protein translocation, *J. Cell Biol.*, 145, 961, 1999.

26. Geissler, A. et al., Biogenesis of the yeast frataxin homolog Yfh1p. Tim44-dependent transfer to mtHSP70 facilitates folding of newly imported proteins into mitochondria, *Eur. J. Biochem.*, 267, 3167, 2000.

27. Westermann, B. et al., Role of the mitochondrial DnaJ homolog Mdj1p as a chaperone for mitochondrially synthesized and imported proteins, *EMBO J.*, 14, 3452, 1995.

28. Cyr, D. M., Coupling chemical energy by the HSP70/tim44 complex to drive protein translocation into mitochondria, *J. Bioenerg. Biomembr.*, 29, 29, 1997.

29. Glick, B. S., Can hsp70 proteins act as force-generating motors?, *Cell*, 80, 11, 1995.

30. Voos, W. et al., Presequence and mature part of preproteins strongly influence the dependence of mitochondrial protein import on heat shock protein 70 in the matrix, *J. Cell Biol.*, 123, 119, 1993.

31. Ungermann, C., Neupert, W., and Cyr, D. M., The role of HSP70 in conferring unidirectionality on protein translocation into mitochondria, *Science*, 266, 1250, 1994.

32. Voos, W. et al., Differential requirement for the mitochondrial HSP70-Tim44 complex in unfolding and translocation of preproteins, *EMBO J.*, 15, 2668, 1996.

33. Luciano, P. and Geli, V., The mitochondrial processing peptidase: function and specificity, *Experientia*, 52, 1077, 1996.

34. Braun, H. P. and Schmitz, U. K., The mitochondrial processing peptidase, *Int. J. Biochem. Cell Biol.*, 29, 1043, 1997.

35. Hawlitschek, G. et al., Mitochondrial protein import: identification of processing peptidase and of PEP, a processing enhancing protein, *Cell*, 53, 795, 1988.

36. Kang, P. J. et al., Requirement for HSP70 in the mitochondrial matrix for translocation and folding of precursor proteins, *Nature*, 348, 137, 1990.

37. Herrmann, J. M. et al., Mitochondrial heat shock protein 70, a molecular chaperone for proteins encoded by mitochondrial DNA, *J. Cell Biol.*, 127, 893, 1994.

38. Horst, M. et al., Sequential action of two HSP70 complexes during protein import into mitochondria, *EMBO J.*, 16, 1842, 1997.

39. Manning-Krieg, U. C., Scherer, P. E., and Schatz, G., Sequential action of mitochondrial chaperones in protein import into the matrix, *EMBO J.*, 10, 3273, 1991.

40. Cheng, M. Y. et al., Mitochondrial heat-shock protein HSP60 is essential for assembly of proteins imported into yeast mitochondria, *Nature*, 337, 620, 1989.

41. Martin, J., Molecular chaperones and mitochondrial protein folding, *J. Bioenerg. Biomembr.*, 29, 35, 1997.

42. Agsteribbe, E. et al., A fatal, systemic mitochondrial disease with decreased mitochondrial enzyme activities, abnormal ultrastructure of the mitochondria and deficiency of heat shock protein 60, *Biochem. Biophys. Res. Commun.*, 193, 146, 1993.

43. Huckriede, A. and Agsteribbe, E., Decreased synthesis and inefficient mitochondrial import of HSP60 in a patient with a mitochondrial encephalomyopathy, *Biochim. Biophys. Acta*, 1227, 200, 1994.

44. Ornatsky, O. I., Connor, M. K., and Hood, D. A., Expression of stress proteins and mitochondrial chaperonins in chronically stimulated skeletal muscle, *Biochem. J.*, 311, 119, 1995.

45. Martinus, R. D. et al., Selective induction of mitochondrial chaperones in response to loss of the mitochondrial genome, *Eur. J. Biochem.*, 240, 98, 1996.

46. Takahashi, M. and Hood, D. A., Chronic stimulation-induced changes in mitochondria and performance in rat skeletal muscle, *J. Appl. Physiol.*, 74, 934, 1993.

47. Takahashi, M. et al., Contractile activity-induced adaptations in the mitochondrial protein import system, *Am. J. Physiol.*, 274, C1380, 1998.

48. Gordon, J. W. et al., Effects of contractile activity on mitochondrial transcription factor A expression in skeletal muscle, *J. Appl. Physiol.*, 90, 389, 2001.

49. Craig, E. E., Chesley, A., and Hood, D. A., Thyroid hormone modifies mitochondrial phenotype by increasing protein import without altering degradation, *Am. J. Physiol.*, 275, C1508, 1998.

50. Schneider, J. J. and Hood, D. A., Effect of thyroid hormone on mtHSP70 expression, mitochondrial import and processing in cardiac muscle, *J. Endocrinol.*, 165, 9, 2000.

51. Carrier, H. et al., Immunolabelling of mitochondrial superoxide dismutase and of HSP60 in muscles harbouring a respiratory chain deficiency, *Neuromuscul. Disord.*, 10, 144, 2000.

52. Rungi, A. A. et al., Events upstream of mitochondrial protein import limit the oxidative capacity of fibroblasts in multiple mitochondrial disease, *Biochim. Biophys. Acta* (in press, 2001).

53. Desjardins, P., de Muys, J. M., and Morais, R., An established avian fibroblast cell line without mitochondrial DNA, *Somat. Cell Mol. Genet.*, 12, 133, 1986.

54. Poulton, J. et al., Deficiency of the human mitochondrial transcription factor h-mtTFA in infantile mitochondrial myopathy is associated with mtDNA depletion, *Hum. Mol. Gen.*, 3, 1763, 1994.

55. Li, K., Neufer, P. D., and Williams, R. S., Nuclear responses to depletion of mitochondrial DNA in human cells, *Am. J. Physiol.*, 269, C1265, 1995.

chapter nine

Gender-specific regulation of HSP70: mechanisms and consequences

Zain Paroo and Earl G. Noble

Contents

I. Gender bias

Males and females are different. Although a seemingly obvious statement, this is a fact too often overlooked in the biomedical research community. The United States Institute of Medicine, a branch of the National Academy of Sciences, recently issued a report[1] compiled over the course of nearly one and a half years by a 16-member panel that arrived at the same conclusion.

In clinical and fundamental research, the oversights can be classified into two categories. The first category involves perpetual male bias. That is, research has almost exclusively investigated the biology of males, and thus the conclusions from these studies are only valid for these models. For example, the development of certain pathologies and the response to some pharmacological treatments are now known to be gender dependent.[1] The second class of errors involve *a priori* assumptions of gender equality and the resulting flaws that follow in terms of experimental design. Scientific literature, particularly that based on human studies, is littered with investigations that group subjects of both sexes. If, for example, the objectives of a study are related to the blood pressure response of a particular pharmaceutical agent, an experimental design grouping together chimpanzees and humans would never receive approval from ethics committees or peer-reviewed publications. Yet historically, this practice has continued with regard to gender.

Recently, however, this oversight has received considerable attention, in part for scientific reasons, as investigators have encountered sexual dimorphism in their respective fields of study. This, combined with political pressures, have prompted various funding agencies to implement directives for gender-specific research. Such initiatives, which have begun to bear fruit, provide tangible incentives for researchers to reverse the discounting of this parameter.

II. Overview of heat shock protein biology

The present review establishes a very specific example of biological sexual dimorphism in the regulation of heat shock protein (HSP) expression with exercise. Heat shock proteins are a class of rapidly inducible products of highly conserved transcriptional units. The induction of HSPs is believed to be mediated through proximal, contiguous, inverted repeats of the sequence nGAAn, termed heat shock elements (HSEs).[2] Induction of the 70-kilodalton stress protein (HSP70) is autoregulated, primarily at the level of transcription.[3,4] In the unstressed cell, HSPs are associated with the primary heat shock transcription factor (HSF1).[5] In response to a variety of stressors, intracellular accumulation of nonnative proteins requires HSP binding to prevent misfolding and aggregation, to facilitate proper refolding, and/or targeting for degradation.[6,7] This activity of HSPs allows cellular HSF1 to acquire HSE-DNA binding and transcriptional activity through trimerization and phosphorylation events, respectively.[5,8] As HSP mRNA is translated, the level of HSPs rises, resulting in sequestration of HSF1.[3] Transcription of the HSP70 gene produces a primary transcript lacking in intervening sequences and with little 5′-untranslated region (UTR) secondary structure.[9] It is these properties of the 2.5-Kb message that are believed to facilitate the preferential synthesis of HSP70 during cellular stress. While HSP70 mRNA is efficiently translated under control and heat shock conditions,[10] the half-life of the relatively unstable transcript increases during periods of stress.[11,12]

III. Exercise as a physiological inducer of HSPs

Locke et al.[13] first reported synthesis of HSPs in a number of rodent tissues following an acute bout of treadmill running. Subsequent to this, a number of other laboratories further characterized this acute response,[14-18] and exercise training models, both short term[19,20] and long term, have since been established.[21-26] However, very little attention has been given to the human biology of HSPs, particularly with respect to exercise models.[27-30]

The precise mechanisms that signal HSP induction with exercise are unknown. However, a number of biological events associated with exercise have independently been shown to induce HSP synthesis in simpler experimental systems, including damage to intracellular structures, increased body temperature, production of reactive oxygen intermediates, ischemia, substrate depletion, decreased pH, loss of calcium homeostasis, reduced intracellular ATP levels, and adrenergic activity.[31] Because many of these putative cellular signals in the HSP response to exercise demonstrate sexual dimorphism,[32,33] it was hypothesized that HSP induction following exercise would also exhibit gender specificity.

IV. Gender-specific response to exercise

The landmark work in the study of gender-specific response to exercise was established by Shumate et al.,[34] where men demonstrated greater levels of circulating creatine kinase (CK, an indicator of exercise-induced muscle damage) than women following a 1-hour bout of stationary cycling at a given relative workload. Of all the possible factors that could underlie this gender-specific response, one of the more obvious differences between males and females is in the content of sex hormones. After establishing this gender-specific response in a rodent model, Amelink, Bar, and co-workers[35,36] provided evidence for the involvement of the female-specific hormone estrogen in this response. Ovariectomized animals exhibited greater serum CK activity than intact females following exercise, and estrogen administration reversed this effect. Komulainen et al.[37] later directly demonstrated gender-specific, exercise-induced skeletal muscle damage through histochemical assessment.

Sex hormones classically exert their effects through interaction with intracellular receptors, with these hormone-receptor complexes acting as transcription factors on target genes.[38] However, tamoxifen, an estrogen receptor antagonist, did not alter the ability of estrogen to protect skeletal muscle in response to damaging electrical stimulation.[39] Moreover, tamoxifen treatment alone attenuated contraction-induced damage. These findings provided evidence against a receptor-mediated mechanism of estrogen and suggested that estrogen and tamoxifen protected skeletal muscle through a common property. The lipophilic characteristics of these compounds have been shown to possess antioxidant potential against lipid peroxidation by reducing membrane fluidity.[40] Thus, the mechanism by which estrogen is believed to protect against exercise-induced tissue damage is through direct stabilization of cellular membranes.[32,33]

V. Gender-specific regulation of HSP70 with exercise

The above work served as the rationale for the present experiments. HSPs are synthesized in response to intracellular accumulation of denatured proteins, which require HSP chaperone activity to maintain cell viability. Thus, the tissue injury occurring with exercise may signal HSP induction. Therefore, females that demonstrate reduced post-exercise injury relative to males should show an attenuated HSP response to exercise.

Indeed, a preliminary study indicated greater skeletal muscle HSP70 content in males than females following exercise.[41] Moreover, estrogen-treated males expressed lower post-exercise HSP70 levels, similar to those observed for females, suggesting that the ovarian hormone is an important factor in this gender-specific response. This pattern of HSP70 expression was also observed in tissues other than skeletal muscle, including liver, lung, and heart, thus implicating a common mechanism of hormone action.

To determine the influence of estrogen on exercise induction of HSP70 in a more discriminating manner, subsequent studies were performed with males, gonadally intact females, and ovariectomized females treated with either placebo or estrogen. The ovaries are the major endogenous source of estrogen, and their removal results in an animal model essentially devoid of significant levels of circulating sex hormones. In skeletal muscle, males demonstrated greater post-exercise HSP70 content than females; removal of the ovaries resulted in HSP induction similar to that observed for males and estrogen treatment to ovariectomized animals reversed this effect.[42] Thus, the sex-specific HSP response to exercise is mediated by the ovarian hormone estrogen. Similar to the series of studies outlined above, this effect was not receptor mediated because tamoxifen treatment in intact females did not alter the response relative to those treated with placebo. Furthermore, 17α-estradiol (a synthetic, receptor inactive stereoisomer) and tamoxifen treatment to ovariectomized (endogenous hormone) animals inhibited induction of HSP70, similar to the effect of 17β-estradiol, further implicating a nongenomic mechanism. That these compounds, which exert different effects at the estrogen receptor, mitigate HSP induction in a fashion similar to that for the endogenous hormone indicates a mechanism of action shared among these compounds. These lipophilic agents have been shown to reduce membrane fluidity and attenuate lipid peroxidation, and have been classified as antioxidant membrane-stabilizing molecules.[40,43] Thus, the mechanism through which these structurally related compounds influence exercise induction of HSP70 is likely through their common indirect antioxidant activity, via physico-chemical maintenance of cell structural integrity. Observations of greater post-exercise induction of the oxidative stress-inducible enzyme HO-1 in placebo-treated ovariectomized animals relative to those treated with estrogen supports this antioxidant membrane-stabilizing mechanism of action.[33,40,44]

It has become increasingly apparent that estrogen influences biological activity far beyond its role in reproduction. Consequently, the hormone has recently become the focus of a myriad of studies over a wide range of scientific disciplines. The most outstanding confound in this, however, has been the use of pharmacological hormone concentrations. The chemical structure of 17β-estradiol contains a phenol ring with a hydroxyl group that is potentially donatable in redox reactions.[45] Similar to the antioxidant vitamin E, for example, this feature of the steroid hormone suggests a possible direct, chain-breaking antioxidant mechanism of action.[46] The micro- and millimolar hormone concentrations employed in many studies would permit biologically significant direct antioxidant activity. However, the picomolar concentrations of circulating estrogen in animals are likely too low to substantially influence cellular reduction-oxidation potentials.

VI. Redox mechanism of HSP induction

In vitro and cell culture work has provided evidence that protein denaturation, and resulting induction of the stress response, are mediated by intracellular redox status. Because a wide variety of chemical inducers of the HSP response have the converging effect of producing protein unfolding, a number of investigators have postulated a more proximal merger. Russo et al.[47] first observed that depletion of cellular GSH (reduced glutathione) resulted in thermotolerance and concomitant synthesis of HSPs. Further characterization of this phenomenon was performed in a number of experimental models using various perturbations by which to manipulate intracellular redox status, demonstrating that oxidation and depletion of non-protein thiols, of which GSH is the most predominant, resulted in HSF1 activation, HSP gene transcription, and HSP synthesis, and maintaining a reducing cellular environment inhibited this response.[48-59] This has led to the hypothesis that protein denaturation, and consequential HSP induction, are signaled via oxidation and/or depletion of intracellular non-protein sulfhydryls (NPSH).[60,61]

This hypothesis was of particular interest to us because estrogen-mediated mitigation of HSP induction might occur through this pathway. The first criterion in this, however, requires the presence of such a mechanism *in vivo*. Physiological inducers of HSF1 activation (exercise and heat shock) were not associated with changes in GSH in either males or females, nor were there any gender differences in redox state following such treatment (unpublished observations). Thus, cellular oxidation is not a required signal for HSP induction in higher-order physiological systems. Therefore, the direct antioxidant activity of estrogen is not the mechanism through which the gender-specific, hormone-mediated exercise induction of HSP70 is mediated.

Furthermore, tamoxifen, which also attenuated HSP70 induction with exercise in ovariectomized animals, does not have a phenolic structure[62] and therefore does not influence biological activities through direct antioxidant action. These observations together support the earlier assertion that the mechanism by which estrogen attenuates HSP70 induction with exercise is likely mediated through indirect antioxidant activity by stabilizing cellular membranes.

VII. Exercise signaling of HSP induction

A number of biological events associated with exercise have independently been shown to induce HSP synthesis. Thus, there are several candidate factors that might signal the HSP response to exercise. The gender-specific, hormone-mediated phenomenon reported here can be used as a model by which this may be resolved. That is, because estrogen inhibits HSP70 synthesis with exercise, those factors that are important in signaling this response are likely mediated by estrogen. The present series of experiments suggests that exercise-induced tissue injury, particularly at cellular membranes, is an important signal for exercise induction of HSP70. Indeed, anative plasma membrane protein conformation has been shown with HSP inducing, non-hyperthermic stress.[61]

For obvious reasons, temperature is the most studied inducer of the HSP response. While body temperature has been shown to mediate the HSP response to exercise,[16,18] the present findings indicate that temperature is not the primary factor in this response. Male rodents, which demonstrated relatively high post-exercise HSP70 content, had higher post-exercise rectal temperatures than intact females and estrogen- and tamoxifen-treated ovariectomized animals, which exhibited relatively low post-exercise HSP70 expression.[42] However, body temperatures among male, tamoxifen-treated intact females, and α-estradiol-treated ovariectomized rodents were not different, despite lower post-exercise HSP70 levels for the latter two groups. Moreover, tamoxifen-treated intact females and α-estradiol-treated ovariectomized rats had significantly higher temperatures than intact females, despite displaying similar post-exercise HSP70 levels. This dissociation of exercising body temperature and HSP70 expression indicates that while temperature may play a role, clearly there are other factors more important in the HSP response to exercise.

VIII. Protective potential of HSPs

Due to the demonstrated cardioprotective potential of HSP70,[63–66] the regulation of this phenomenon in the heart and its physiological outcomes were of particular interest. The proposed use of HSPs in clinical settings began with observations that induction of stress proteins via sublethal heat exposure conferred protection against subsequent lethal heat shock, a phenomenon termed "thermotolerance." Because the mechanism of this protection was believed to involve a general stabilization of cellular proteins, investigators tested the hypothesis that heat-induced HSP expression would enhance defense against other forms of stress. Indeed, these studies demonstrated the ability of HSPs to protect against a variety of otherwise lethal stressors.[68,69]

Among the first reports of this "cross-tolerance" effect of stress proteins in the heart, Currie et al.[70] demonstrated that heat shock enhanced ventricular recovery following ischemia-reperfusion, and later showed a close temporal relationship between the kinetics of heat-induced HSP70 accumulation and functional recovery from myocardial ischemia-reperfusion.[71] A number of groups further demonstrated the cardioprotective effect of HSP70 *in vivo*.[72-74] Direct evidence for HSP70-mediated protection against ischemic injury is provided by studies employing transgenic and transfected murine and rodent hearts, where those overexpressing HSP70 displayed increased tolerance to myocardial trauma.[63-66] While these perturbations have been useful in demonstrating the potential benefits of HSPs, the use of HSP-mediated cardioprotection in medical settings requires interventions with greater physiological relevance.

IX. Functional outcomes of the gender-specific HSP response to exercise

Similar to our observations in skeletal muscle, following exercise, males demonstrated greater cardiac HSP70 than gonadally intact females.[75] Removal of the ovaries resulted in HSP induction similar to that observed for males, and estrogen treatment to ovariectomized animals reversed this effect. Further, the gender-specific, hormone-mediated induction of cardiac HSP70 is likely regulated at the level of transcription, as male and placebo-treated ovariectomized rodents also demonstrated greater levels of myocardial HSP70 mRNA and HSF1-HSE DNA binding than intact and estrogen-treated ovariectomized females. Moreover, this attenuated signaling was likely the result of non-receptor-mediated membrane stabilization because 17α-estradiol or tamoxifen treatment to ovariectomized animals attenuated the response in a manner similar to that for 17β-estradiol. That these findings were identical to those observed in skeletal muscle affirms a common mechanism of hormone action postulated from preliminary data.

To determine the physiological importance of this gender-specific, hormone-mediated HSP response, cardiac function following ischemia-reperfusion was assessed in control and exercised male, intact female, and ovariectomized female rodents. Exercise improved postischemic left ventricular developed pressure, maximal rate of contraction, and maximal rate of relaxation, and reduced left ventricular end-diastolic pressure in males.[75] Intact females demonstrated no such effect with exercise. However, exercise improved all parameters of function following myocardial ischemia-reperfusion in ovariectomized females, similar to what was observed in males. These functional outcomes were related to myocardial structural integrity as exercised male and ovariectomized animals demonstrated lower creatine kinase (CK) efflux following ischemia relative to their corresponding control and female groups. These findings are in line with the putative maintenance of cellular structure conferred by HSPs. In summary, male and ovariectomized animals, which induced HSP70 substantially with exercise, demonstrated greater recovery of myocardial function following ischemia than intact females, which exhibited a marginal stress response.

X. Manipulation of HSP70 expression

To determine a causal relationship between the improved recovery of myocardial function with exercise and exercise induction of HSP70, manipulation of HSP70 expression was achieved through an antisense oligonucleotide approach. Antisense oligonucleotides are short, synthetic strands of DNA designed complementary to a particular segment of the transcript of interest in order to inhibit its expression.[76] Treatment of exercised males with antisense oligonucleotides designed against the translational start site of inducible HSP70 transcripts[77] knocked out the HSP response to exercise and mitigated the improvement in recovery of postischemic cardiac function.[75] This direct and specific modulation of HSP70 expression provides evidence that the protective effects of exercise on cardiac function are at least, in part, HSP70 dependent. Thus, the gender-specific, hormone-mediated induction of HSP70 with exercise results in cardioprotective adaptation in males but not in females.

XI. Significance and impact

Estrogen has generally been believed to be a protective factor in the onset and effects of cardiovascular disease.[38] This is largely based on epidemiological findings of lower incidence of cardiovascular disease among women relative to men until menopause, after which this disparity diminishes.[78–80] Further, post-menopausal women receiving estrogen replacement therapy are found to have lower rates of coronary heart disease.[81,82] Moreover, a vast amount of experimental work has characterized potential biological mechanisms for these observations.[83–85] This association between estrogen and heart health, however, is the subject of increasing controversy. For example, the disparity in incidence of heart disease between men and women is reduced when lifestyle factors are accounted for.[86] Furthermore, recent randomized clinical trials to determine the influence of hormone replacement therapy on coronary heart disease have yielded equivocal results.[87,88] Moreover, the preponderance of fundamental research suggesting positive cardiovascular outcomes with estrogen treatment are derived mainly from studies performed with supraphysiological experimental hormone concentrations.

Clinical data indicate possible severe negative outcomes of estrogen as pre-menopausal women, following myocardial infarction, demonstrate greater rates of mortality relative to males.[89–91] This phenomenon has been outlined experimentally where estrogen-treated ovariectomized rodents displayed larger infarct size relative to those placebo treated.[92] Thus, among cardiovascular clinicians and researchers, the association between estrogen and heart health is being revisited.

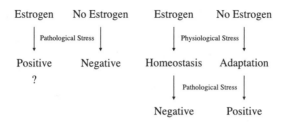

Figure 9.1 Schematic illustrating reversal fo the relative protective effects of estrogen when severe stress is preceded by sublethal physiological stimuli.

The present series of experiments indicate that the effects of estrogen can be interpreted as either advantageous or disadvantageous. Because HSP70 is stress inducible, attenuated HSP70 levels following exercise in estrogen-positive relative to estrogen-naïve animals suggest a beneficial hormonal effect. Indeed, a membrane-stabilizing mechanism of action would aid in maintaining normal cellular function with exercise stress.[62] However, maintenance of homeostasis with physiological stress reduces signaling of adaptive processes. Thus, upon subsequent severe stimuli, this apparent protective action of estrogen manifests as a disadvantage. That is, although male and ovariectomized females demonstrated high cellular stress with exercise (as per the relatively high induction of HSP70), this stress signaled protective systems mitigating the effects of subsequent severe ischemia-reperfusion stress. Gonadally intact females, which displayed relatively low levels of cellular stress with exercise (as per relatively low HSP70 expression), demonstrated increased susceptibility to ensuing ischemia-reperfusion (Figure 9.1).

Although widely believed to exert cardioprotective effects, exercise is one of the most often, yet most poorly, prescribed treatments in medicine. A generic term clinically, exercise prescription generally lacks reference to mode, intensity, and/or duration. The precise manner in which exercise improves cardiovascular disease profile has not been fully defined. The present work supports direct involvement of HSP70 in exercise-conferred cardioprotection. However, a whole-body perturbation such as treadmill running undoubtedly influences a number of endogenous defense systems. Further, direct evidence for exercise modulation of other protective mechanisms has been reported.[93] In light of the present work, it is of interest if such endogenous defense mechanisms also demonstrate gender dimorphism. Further still is the question of whether there are other physiological stimuli that confer protection and whether similar sex and hormone mediation of these exist. The novel, gender-specific, hormone-mediated HSP response described in this review suggests that exercise might be more important for males than females in defending against the effects of cardiovascular disease and offers a mechanism by which males might reduce the gender gap in susceptibility to adverse cardiac events.

XII. Limitations and future directions

The outstanding question the present work poses relates to its significance for humans. Sexual dimorphism with regard to therapeutic effects of exercise in coronary heart disease patients has been reported.[94–96] Further, although pre-menopausal women demonstrate a lower incidence of cardiovascular disease than age-matched men, acute cardiac trauma results in higher rates of mortality and poorer patient prognosis in women than men.[89–91] However, the factors underlying such disparities are unknown. The present work provides a starting point from which these questions can be addressed. Estrogen is generally viewed as a protective agent, particularly for its putative positive effects on heart health. However, this is largely based on observational data and nonphysiological experimental models.[87,97,98] The present work revisits this notion. Because HSP70 is stress inducible, the lower post-exercise expression of HSP70 in females relative to males suggests a protective hormonal effect. However, this protection, ameliorates the induction of adaptive mechanisms that ultimately enhance an organismal defense. This process appears as a pattern throughout biology and, indeed, in nonbiological models, such that long-term benefits require short-term costs.

Physiological adaptation requires some form of injury in order for remodeling processes to take place.[30] As such, attenuating these pathways, although seemingly protective initially, results in longer term disadvantage. Indeed, a promising approach in the treatment of pathologies of the central nervous system, which does not demonstrate an "injury" response, is to introduce these factors exogenously to damaged areas.[99] The current work provides a holistic view of this paradigm in a physiological setting and suggests that similar general principles apply to gender disparities in the human population and introduces the possibility of stress-based approaches in the development of cardiac therapies.

One of the critical deficiencies in the HSP area—and not addressed in the present chapter—is the regulation of these transcriptional units in humans. While exercise has been shown to induce the stress response in humans, the human biology of HSPs has been shown to differ from that of lower mammals, particularly with respect to the relatively high constitutive levels of expression which may require more severe or more frequent stimuli to cause induction of the response in humans.[100–102] A major obstacle in this area involves obtaining healthy tissue samples. However, a unique population of healthy human hearts has been identified in organ donors who have expired due to noncardiovascular causes of death.[103] Such models will be of tremendous use in future examinations of human HSP regulation.

The acute nature of the experiments reported herein are an additional limitation, and future studies employing chronic exercise training will underscore the present gender-specific adaptations to exercise. Such acute experiments, however, are necessary in elucidating the mechanisms that mediate phenomena observed in chronic studies. For example, following long-term exercise training, male and ovariectomized rodents demonstrate greater post-ischemic cardiac function relative to gonadally intact females (Dr. Russ Moore, personal communication). In addition, male rodents, in response to long-term exercise training, have also been shown to demonstrate improved basal cardiac function, while females exhibit no such effect.[104–107] The length of such studies imposes significant challenges in elucidating the precise mechanisms that underlie these findings. Thus, the results of the present experiments employing acute models may be helpful in this regard.

XIII. Summary

The findings of gender-specific regulation of HSP expression reviewed here address important shortcomings in the field through the use of physiological models relevant to the biology of males and females. Further, the present work serves as a direct and comprehensive example of how whole-body perturbations influence molecular activities, which in turn regulate physiological processes. In the HSP area, this relationship between the regulation of gene expression and functional outcomes deserves increasing attention to underscore the importance of these tremendously important biological players.

References

1. Wizemann, T. M. and Pardue, M. L., Exploring the biological contributions to human health. Does sex matter?, Report by committee on understanding the biology of sex and gender differences, National Academy Press, Washington, D.C., 2001.
2. Kingston, R. E., Schuetz, T. J., and Larin, Z., Heat-inducible human factor that binds to a human HSP70 promoter, *Mol. Cell. Biol.*, 7, 1530, 1987.
3. Kim, D., Ouyang, H., and Li, G. C., Heat shock protein HSP70 accelerates the recovery of heat-shocked mammalian cells through its modulation of heat shock transcription factor HSF1, *Proc. Natl. Acad. Sci. U.S.A.*, 92, 2126, 1995.
4. Shi, Y., Mosser, D. D., and Morimoto, R. I., Molecular chaperones as HSF1-specific transcriptional repressors, *Genes Dev.*, 12, 654, 1998.
5. Abravaya, K. et al., The human heat shock protein HSP70 interacts with HSF, the transcription factor that regulates heat shock gene expression, *Genes Dev.*, 6, 1153, 1992.
6. Hochstrasser, M., Ubiquitin and intracellular protein degradation, *Curr. Opin. Cell. Biol.*, 4, 1024, 1992.
7. Ellis, R. J. and Hartl, F. U., Protein folding in the cell: competing models of chaperonin function, *FASEB J.*, 10, 20, 1996.
8. Mosser, D. D., Duchaine, J., and Massie, B., The DNA-binding activity of the human heat shock transcription factor is regulated *in vivo* by HSP70, *Mol. Cell. Biol.*, 13, 5427, 1993.
9. Hess, M. A. and Duncan, R. F., Sequence and structure determinants of *Drosophila* HSP70 mRNA translation: 5′UTR secondary structure specifically inhibits heat shock protein mRNA translation, *Nucl. Acids Res.*, 24, 2441, 1996.
10. Hickey, E. D. and Weber, L. A., Modulation of heat-shock polypeptide synthesis in HeLa cells during hyperthermia and recovery, *Biochemistry*, 21, 1513, 1982.

11. Theodorakis, N. G. and Morimoto, R. I., Posttranscriptional regulation of HSP70 expression in human cells: effects of heat shock, inhibition of protein synthesis, and adenovirus infection on translation and mRNA stability, *Mol. Cell. Biol.*, 7, 4357, 1987.

12. DiDomenico, B. J., Bugaisky, G. E., and Lindquist, S., The heat shock response is self-regulated at both the transcriptional and posttranscriptional levels, *Cell*, 31, 593, 1982.

13. Locke, M., Noble, E. G., and Atkinson, B. G., Exercising mammals synthesize stress proteins, *Am. J. Physiol.*, 258, C723, 1990.

14. Salo, D. C., Donovan, C. M., and Davies, K. J., HSP70 and other possible heat shock or oxidative stress proteins are induced in skeletal muscle, heart, and liver during exercise, *Free Radic. Biol. Med.*, 11, 239, 1991.

15. Chen, H. W. et al., Synthesis of HSP72 induced by exercise in high temperature, *Chin. J. Physiol.*, 38, 241, 1995.

16. Skidmore, R. et al., HSP70 induction during exercise and heat stress in rats: role of internal temperature, *Am. J. Physiol.*, 268, R92, 1995.

17. Hernando, R. and Manso, R., Muscle fibre stress in response to exercise: synthesis, accumulation and isoform transitions of 70-kDa heat-shock proteins, *Eur. J. Biochem.*, 243, 460, 1997.

18. Taylor, R. P., Harris, M. B., and Starnes, J. W., Acute exercise can improve cardioprotection without increasing heat shock protein content, *Am. J. Physiol.*, 276, H1098, 1999.

19. Locke, M. et al., Enhanced postischemic myocardial recovery following exercise induction of HSP 72, *Am. J. Physiol.*, 269, H320, 1995.

20. Kelly, D. A. et al., Effect of vitamin E deprivation and exercise training on induction of HSP70, *J. Appl. Physiol.*, 81, 2379, 1996.

21. Demirel, H. A. et al., Exercise training reduces myocardial lipid peroxidation following short-term ischemia-reperfusion, *Med. Sci. Sports Exerc.*, 30, 1211, 1998.

22. Powers, S. K. et al., Exercise training improves myocardial tolerance to *in vivo* ischemia-reperfusion in the rat, *Am. J. Physiol.*, 275, R1468, 1998.

23. Demirel, H. A. et al., The effects of exercise duration on adrenal HSP72/73 induction in rats, *Acta Physiol. Scand.*, 167, 227, 1999.

24. Noble, E. G. et al., Differential expression of stress proteins in rat myocardium after free wheel or treadmill run training, *J. Appl. Physiol.*, 86, 1696, 1999.

25. Gonzalez, B., Hernando, R., and Manso, R., Stress proteins of 70 kDa in chronically exercised skeletal muscle, *Pflügers Arch.*, 440, 42, 2000.

26. Harris, M. B. and Starnes, J. W., Effects of body temperature during exercise training on myocardial adaptations, *Am. J. Physiol. Heart Circ. Physiol.*, 280, H2271, 2001.

27. Puntschart, A. et al., HSP70 expression in human skeletal muscle after exercise, *Acta Physiol. Scand.*, 157, 411, 1996.

28. Liu, Y. et al., Human skeletal muscle HSP70 response to training in highly trained rowers, *J. Appl. Physiol.*, 86, 101, 1999.

29. Febbraio, M. A. and Koukoulas, I., HSP72 gene expression progressively increases in human skeletal muscle during prolonged, exhaustive exercise, *J. Appl. Physiol.*, 89, 1055, 2000.

30. Khassaf, M. et al., Time course of responses of human skeletal muscle to oxidative stress induced by nondamaging exercise, *J. Appl. Physiol.*, 90, 1031, 2001.

31. Locke, M., The cellular stress response to exercise: role of stress proteins, *Exerc. Sport Sci. Rev.*, 25, 105, 1997.

32. Tiidus, P. M., Can estrogens diminish exercise induced muscle damage?, *Can. J. Appl. Physiol.*, 20, 26, 1995.

33. Bar, P. R. and Amelink, G. J., Protection against muscle damage exerted by oestrogen: hormonal or antioxidant action?, *Biochem. Soc. Trans.*, 25, 50, 1997.

34. Shumate, J. B. et al., Increased serum creatine kinase after exercise: a sex-linked phenomenon, *Neurology*, 29, 902, 1979.

35. Amelink, G. J. and Bar, P. R., Exercise-induced muscle protein leakage in the rat. Effects of hormonal manipulation, *J. Neurol. Sci.*, 76, 61, 1986.

36. Bar, P. R. et al., Prevention of exercise-induced muscle membrane damage by oestradiol, *Life Sci.*, 42, 2677, 1988.

37. Komulainen, J. et al., Gender differences in skeletal muscle fibre damage after eccentrically biased downhill running in rats, *Acta Physiol. Scand.*, 165, 57, 1999.

38. Mendelsohn, M. E. and Karas, R. H., The protective effects of estrogen on the cardiovascular system, *N. Engl. J. Med.*, 340, 1801, 1999.

39. Koot, R. W. et al., Tamoxifen and oestrogen both protect the rat muscle against physiological damage, *J. Steroid Biochem. Mol. Biol.*, 40, 689, 1991.

40. Wiseman, H., Quinn, P., and Halliwell, B., Tamoxifen and related compounds decrease membrane fluidity in liposomes. Mechanism for the antioxidant action of tamoxifen and relevance to its anticancer and cardioprotective actions?, *FEBS Lett.*, 330, 53, 1993.

41. Paroo, Z., Tiidus, P. M., and Noble, E. G., Estrogen attenuates HSP 72 expression in acutely exercised male rodents, *Eur. J. Appl. Physiol.*, 80, 180, 1999.

42. Paroo, Z., Dipchand, E. S., and Noble, E. G., Estrogen attenuates post-exercise HSP70 induction in skeletal muscle, submitted for publication, 2001.

43. Behl, C. et al., Neuroprotection against oxidative stress by estrogens: structure-activity relationship, *Mol. Pharmacol.*, 51, 535, 1997.

44. Shibahara, S. et al., Cloning and expression of cDNA for rat heme oxygenase, *Proc. Natl. Acad. Sci. U.S.A.*, 82, 7865, 1985.

45. Sugioka, K., Shimosegawa, Y., and Nakano, M., Estrogens as natural antioxidants of membrane phospholipid peroxidation, *FEBS Lett.*, 210, 37, 1987.

46. Subbiah, M. T. et al., Antioxidant potential of specific estrogens on lipid peroxidation, *J. Clin. Endocrinol. Metab.*, 77, 1095, 1993.

47. Russo, A., Mitchell, J. B., and McPherson, S., The effects of glutathione depletion on thermotolerance and heat stress protein synthesis, *Br. J. Cancer*, 49, 753, 1984.

48. Freeman, M. L. et al., Diamide exposure, thermal resistance, and synthesis of stress (heat shock) proteins, *Biochem. Pharmacol.*, 36, 21, 1987.

49. Freeman, M. L., Meredith, M. J., and Laszlo, A., Depletion of glutathione, heat shock protein synthesis, and the development of thermotolerance in Chinese hamster ovary cells, *Cancer Res.*, 48, 7033, 1988.

50. Freeman, M. L. et al., Synthesis of HSP-70 is enhanced in glutathione-depleted Hep G2 cells, *Radiat. Res.*, 135, 387, 1993.

51. Huang, L. E. et al., Thiol reducing reagents inhibit the heat shock response. Involvement of a redox mechanism in the heat shock signal transduction pathway, *J. Biol. Chem.*, 269, 30718, 1994.

52. Sierra-Rivera, E. et al., Synthesis of heat shock proteins following oxidative challenge: role of glutathione, *Int. J. Hyperthermia*, 10, 573, 1994.

53. Freeman, M. L. et al., Characterization of a signal generated by oxidation of protein thiols that activates the heat shock transcription factor, *J. Cell. Physiol.*, 164, 356, 1995.

54. Jacquier-Sarlin, M. R. and Polla, B. S., Dual regulation of heat-shock transcription factor (HSF) activation and DNA-binding activity by H_2O_2: role of thioredoxin, *Biochem. J.*, 318 (Pt. 1), 187, 1996.

55. Liu, H., Lightfoot, R., and Stevens, J. L., Activation of heat shock factor by alkylating agents is triggered by glutathione depletion and oxidation of protein thiols, *J. Biol. Chem.*, 271, 4805, 1996.

56. Senisterra, G. A. et al., Destabilization of the Ca^{2+}-ATPase of sarcoplasmic reticulum by thiol-specific, heat shock inducers results in thermal denaturation at 37 degrees C, *Biochemistry*, 36, 11002, 1997.

57. McDuffee, A. T. et al., Proteins containing non-native disulfide bonds generated by oxidative stress can act as signals for the induction of the heat shock response, *J. Cell. Physiol.*, 171, 143, 1997.

58. Zou, J. et al., Correlation between glutathione oxidation and trimerization of heat shock factor 1, an early step in stress induction of the HSP response, *Cell Stress Chaperones*, 3, 130, 1998.

59. Hatayama, T. and Hayakawa, M., Differential temperature dependency of chemical stressors in HSF1-mediated stress response in mammalian cells, *Biochem. Biophys. Res. Commun.*, 265, 763, 1999.

60. Freeman, M. L. et al., Destabilization and denaturation of cellular protein by glutathione depletion, *Cell Stress Chaperones*, 2, 191, 1997.

61. Freeman, M. L. et al., On the path to the heat shock response: destabilization and formation of partially folded protein intermediates, a consequence of protein thiol modification, *Free Radic. Biol. Med.*, 26, 737, 1999.

62. Wiseman, H., Tamoxifen: new membrane-mediated mechanisms of action and therapeutic advances, *Trends Pharmacol. Sci.*, 15, 83, 1994.

63. Suzuki, K. et al., *In vivo* gene transfection with heat shock protein 70 enhances myocardial tolerance to ischemia-reperfusion injury in rat, *J. Clin. Invest.*, 99, 1645, 1997.

64. Trost, S. U. et al., Protection against myocardial dysfunction after a brief ischemic period in transgenic mice expressing inducible heat shock protein 70, *J. Clin. Invest.*, 101, 855, 1998.

65. Plumier, J. C. et al., Transgenic mice expressing the human heat shock protein 70 have improved post-ischemic myocardial recovery, *J. Clin. Invest.*, 95, 1854, 1995.

66. Marber, M. S. et al., Overexpression of the rat inducible 70-kD heat stress protein in a transgenic mouse increases the resistance of the heart to ischemic injury, *J. Clin. Invest.*, 95, 1446, 1995.

67. Li, G. C. and Werb, Z., Correlation between synthesis of heat shock proteins and development of thermotolerance in Chinese hamster fibroblasts, *Proc. Natl. Acad. Sci. U.S.A.*, 79, 3218, 1982.

68. Lee, K. J. and Hahn, G. M., Abnormal proteins as the trigger for the induction of stress responses: heat, diamide, and sodium arsenite, *J. Cell Physiol.*, 136, 411, 1988.

69. Nayeem, M. A. et al., Delayed preconditioning of cultured adult rat cardiac myocytes: role of 70- and 90-kDa heat stress proteins, *Am. J. Physiol.*, 273, H861, 1997.

70. Currie, R. W. et al., Heat-shock response is associated with enhanced postischemic ventricular recovery, *Circ. Res.*, 63, 543, 1988.

71. Karmazyn, M., Mailer, K., and Currie, R. W., Acquisition and decay of heat-shock-enhanced postischemic ventricular recovery, *Am. J. Physiol.*, 259, H424, 1990.

72. Marber, M. S. et al., Cardiac stress protein elevation 24 hours after brief ischemia or heat stress is associated with resistance to myocardial infarction, *Circulation,* 88, 1264, 1993.

73. Currie, R. W., Tanguay, R. M., and Kingma, J. G., Jr., Heat-shock response and limitation of tissue necrosis during occlusion/reperfusion in rabbit hearts [see Comments], *Circulation,* 87, 963, 1993.

74. Hutter, M. M. et al., Heat-shock protein induction in rat hearts. A direct correlation between the amount of heat-shock protein induced and the degree of myocardial protection, *Circulation,* 89, 355, 1994.

75. Paroo, Z. et al., Exercise improves post-ischemic cardiac function in males but not females, *Am. J. Physiol.,* 282, C245, 2002.

76. Hogrefe, R. I., An antisense oligonucleotide primer, *Antisense Nucl. Acid Drug Dev.,* 9, 351, 1999.

77. Khanna, A., Aten, R. F., and Behrman, H. R., Heat shock protein-70 induction mediates luteal regression in the rat, *Mol. Endocrinol.,* 9, 1431, 1995.

78. Godsland, I. F. et al., Sex, plasma lipoproteins, and atherosclerosis: prevailing assumptions and outstanding questions, *Am. Heart J.,* 114, 1467, 1987.

79. Isles, C. G. et al., Relation between coronary risk and coronary mortality in women of the Renfrew and Paisley survey: comparison with men [see Comments], *Lancet,* 339, 702, 1992.

80. Kostis, J. B. et al., Sex differences in the management and long-term outcome of acute myocardial infarction. A statewide study. MIDAS Study Group. Myocardial Infarction Data Acquisition System, *Circulation,* 90, 1715, 1994.

81. Stampfer, M. J. et al., Postmenopausal estrogen therapy and cardiovascular disease. Ten-year follow-up from the nurses' health study, *N. Engl. J. Med.,* 325, 756, 1991.

82. Grady, D. et al., Hormone therapy to prevent disease and prolong life in post-menopausal women [see Comments], *Ann. Intern. Med.,* 117, 1016, 1992.

83. Williams, J. K., Adams, M. R., and Klopfenstein, H. S., Estrogen modulates responses of atherosclerotic coronary arteries, *Circulation,* 81, 1680, 1990.

84. Delyani, J. A. et al., Protection from myocardial reperfusion injury by acute administration of 17 beta-estradiol, *J. Mol. Cell. Cardiol.,* 28, 1001, 1996.

85. Kim, Y. D. et al., 17-beta Estradiol regulation of myocardial glutathione and its role in protection against myocardial stunning in dogs, *J. Cardiovasc. Pharmacol.,* 32, 457, 1998.

86. Haynes, S. G. and Feinleib, M., Women, work and coronary heart disease: prospective findings from the Framingham heart study, *Am. J. Public Health,* 70, 133, 1980.

87. Hulley, S. et al., Randomized trial of estrogen plus progestin for secondary prevention of coronary heart disease in postmenopausal women. Heart and Estrogen/ progestin Replacement Study (HERS) Research Group, *J. Am. Med. Assoc.,* 280, 605, 1998.

88. Harrington, R. D. and Hooton, T. M., Urinary tract infection risk factors and gender, *J. Gend. Specif. Med.,* 3, 27, 2000.

89. Greenland, P. et al., In-hospital and 1-year mortality in 1,524 women after myocardial infarction. Comparison with 4,315 men, *Circulation,* 83, 484, 1991.

90. Malacrida, R. et al., A comparison of the early outcome of acute myocardial infarction in women and men. The Third International Study of Infarct Survival Collaborative Group, *N. Engl. J. Med.,* 338, 8, 1998.

91. Vaccarino, V. et al., Sex-based differences in early mortality after myocardial infarction. National Registry of Myocardial Infarction 2 Participants, *N. Engl. J. Med.*, 341, 217, 1999.

92. Smith, P. J. et al., Effects of estrogen replacement on infarct size, cardiac remodeling, and the endothelin system after myocardial infarction in ovariectomized rats, *Circulation*, 102, 2983, 2000.

93. Yamashita, N. et al., Exercise provides direct biphasic cardioprotection via manganese superoxide dismutase activation, *J. Exp. Med.*, 189, 1699, 1999.

94. Kohrt, W. M. et al., Effects of gender, age, and fitness level on response of VO2max to training in 60-71 yr olds, *J. Appl. Physiol.*, 71, 2004, 1991.

95. Spina, R. J. et al., Differences in cardiovascular adaptations to endurance exercise training between older men and women, *J. Appl. Physiol.*, 75, 849, 1993.

96. Spina, R. J. et al., Gender-related differences in left ventricular filling dynamics in older subjects after endurance exercise training, *J. Gerontol. A, Biol. Sci. Med. Sci.*, 51, B232, 1996.

97. Kim, Y. D. et al., 17 beta-Estradiol prevents dysfunction of canine coronary endothelium and myocardium and reperfusion arrhythmias after brief ischemia/reperfusion, *Circulation*, 94, 2901, 1996.

98. Herrington, D. M. et al., Effects of estrogen replacement on the progression of coronary-artery atherosclerosis, *N. Engl. J. Med.*, 343, 522, 2000.

99. Schwartz, M. et al., Potential repair of rat spinal cord injuries using stimulated homologous macrophages, *Neurosurgery*, 44, 1041, 1999.

100. McGrath, L. B. and Locke, M., Myocardial self-preservation: absence of heat shock factor activation and heat shock proteins 70 m accumulation in the human heart during cardiac surgery, *J. Card. Surg.*, 10, 400, 1995.

101. McGrath, L. B. et al., Heat shock protein (HSP 72) expression in patients undergoing cardiac operations, *J. Thorac. Cardiovasc. Surg.*, 109, 370, 1995.

102. Locke, M., Tanguay, R. M., and Ianuzzo, C. D., Constitutive expression of HSP 72 in swine heart, *J. Mol. Cell. Cardiol.*, 28, 467, 1996.

103. Knowlton, A. A. et al., Differential expression of heat shock proteins in normal and failing human hearts, *J. Mol. Cell. Cardiol.*, 30, 811, 1998.

104. Schaible, T. F. and Scheuer, J., Cardiac function in hypertrophied hearts from chronically exercised female rats, *J. Appl. Physiol.*, 50, 1140, 1981.

105. Schaible, T. F., Penpargkul, S., and Scheuer, J., Cardiac responses to exercise training in male and female rats, *J. Appl. Physiol.*, 50, 112, 1981.

106. Schaible, T. F. and Scheuer, J., Effects of physical training by running or swimming on ventricular performance of rat hearts, *J. Appl. Physiol.*, 46, 854, 1979.

107. Malhotra, A., Buttrick, P., and Scheuer, J., Effects of sex hormones on development of physiological and pathological cardiac hypertrophy in male and female rats, *Am. J. Physiol.*, 259, H866, 1990.

chapter ten

The role of heat shock proteins in modulating the immune response

Pope L. Moseley

Contents

I. Introduction

The heat shock proteins (HSPs) are generated by all living organisms. While first identified as gene products of cells exposed to thermal stress, the HSPs are essential for the normal function of nonstressed cells because of their roles as protein/peptide chaperones. This peptide/protein transport function has been implicated in activities as diverse as receptor movement and regulation and in mitochondrial biogenesis. In addition, the constitutively expressed member of the HSP70 family, HSC70, is involved in the management of peptides during the process of protein synthesis.

The HSP genes undergo dramatic and rapid activation in response to an ever-increasing list of stresses. This gene activation and subsequent HSP protein accumulation following a stress stimulus are associated with a state of relative tolerance to subsequent, otherwise lethal, levels of stress. In single-cell systems, HSP accumulation is associated with survival of physical and toxicological stresses such as hyperthermia, hyperoxia, and heavy metals, among others. In the intact organism, HSP accumulation is also associated with survival in response to moderate levels of thermal stress, ischemia-reperfusion injury, and to the cytotoxic effects of inflammatory cytokines such as tumor necrosis factor-alpha (TNF-α) and interleukin-1 (IL-1).

If a causative link between cellular HSP accumulation and subsequent tolerance to stress exists, then HSP gene activation can be seen as the default pathway by which an individual cell or single-cell organism responds to the stress stimulus. This default pathway has been preserved over broad evolutionary distances and exists in all living things — from bacteria to man. In fact, the association of HSP accumulation with tolerance of target cells to cytokines such as TNF-α and IL-1 is an interesting variation on the theme of a primitive, single-cell response to stress. While the HSPs have been understood in their roles as intracellular transporters and peptide/protein managers, the recent literature has developed and demonstrated a role for the HSPs in the extracellular microenvironment in activating a multifaceted immune/inflammatory response involving cytokine release and the activation of a variety of immune effector cells.

The goals of this chapter are threefold: (1) to examine issues regarding the consequences of the activation of the HSP response — the single cell's default pathway to address environmental stresses on the function of multicellular systems; (2) to examine the seeming dichotomy between HSPs in the intracellular environment associated with inflammatory tolerance and in the extracellular milieu as immune/inflammatory activators; and (3) to examine a paradigm to reconcile this dichotomy.

II. Impact of HSP regulation in multicellular systems

One of the most complex issues regarding HSPs relates to their role in multicellular systems. In both single-cell and multicellular systems, HSPs have been demonstrated to serve similar roles in peptide transport and protein translation. Their role in cellular adaptation to environmental and other stresses has been documented elsewhere and is based on studies using both conditioning stimuli such as heat and, more recently, gene overexpression. The body of evidence that supports the role of HSPs in cellular tolerance is based on a series of premises. First, the stresses that cause HSP accumulation also result in thermotolerance. Second, both the induction and decay of thermotolerance parallel the accumulation and degradation of HSPs. Third, cellular manipulations that block HSP accumulation or that result in the overexpression of certain HSPs have been shown to either increase or decrease heat sensitivity, respectively. In the *Drosophila* cell culture model, for example, cells with increased copies of HSP70 genes show enhanced thermotolerance.[1] At the same time, the HSP70 homologue in a bacterial model preserves RNA polymerase function during hyperthermia, yet deletion mutants have only a small decrease in heat tolerance.[2,3]

One aspect of the HSP response is the effect of its induction on other aspects of cellular function. Following the recovery from stress-induced translational arrest, the HSP message is translated to the exclusion of other messages. This preferential translation of HSPs negatively impacts cell growth and other functions of the nonstressed cell. The end result is that cells undergoing a stress response and HSP synthesis will be at an advantage in terms of stress survival but at a disadvantage in terms of cell growth or cell signal transduction. The impact of this alteration in normal cell function may have important consequences in multicellular systems, which rely on constant cell-to-cell interactions. Interestingly, there is little data on the negative impact of a stress response in multicellular systems. While studies evaluating the functional impact of HSP accumulation are difficult in the intact organism, experimental models using *in vitro* multicellular systems such as reconstituted monolayers may provide insight into the role of the stress response in individual cells and the consequence of this response for the function of the multicellular system. Using nonlethal hyperthermia as a stimulus, *in vitro* epithelial monolayers composed of Madin-Darby canine kidney epithelial cells show reversible alterations in electrical resistance. Heat conditioning sufficient to result in HSP accumulation in the cells is associated with an attenuation of this permeability change to a subsequent hyperthermic stress. Because this stress stimulus is reversible and nonlethal, these studies suggest that HSP accumulation may not only be associated with cell survival as defined by classical thermotolerance, but also with a state of physiologic thermotolerance where a multicellular system maintains its cell-to-cell cooperative function in the face of hyperthermia. Given previous studies in single cells implicating the HSPs directly in thermotolerance, it is conceivable that this HSP accumulation in a multicellular system is responsible for the maintenance of epithelial barrier function. At the same time, the preservation of epithelial function associated with HSP accumulation might be equally well-explained by the decreased requirement of the cells of the monolayer for HSP accumulation during the subsequent heat stress. This latter explanation relies on the fact that HSP accumulation is titrated to the degree of stress, applied to cells. Thus, cells undergoing a conditioning stress generate a stress response with cellular HSP accumulation. Under the subsequent stress, a lesser cellular stress response is invoked compared to naïve cells, allowing cells of the monolayer to continue to function as a unit without resorting to the HSP default pathway associated with individual cell survival.

As discussed in the next section, HSP-associated tolerance occurs not only with environmental stresses but also with stresses that are particular to multicellular systems. The association of the stress response with tolerance of target cells to cytokines and the modulation of cytokine production by immune effector cells represent an interesting adaptive aspect of the HSP response to the transduction of signals generated in higher eukaryotes.

III. HSPs in the immune response

A. HSPs in the intracellular environment

The involvement of HSPs within cells in the immune response can be divided into two sets of effects: (1) the effect of HSPs on cells exposed to cytokine stress, and (2) the role of HSPs in cytokine production by immune effector cells.

The observation that heat conditioning with subsequent HSP accumulation could be important in down-regulating the inflammatory response was demonstrated in a whole-animal model of endotoxin shock.[4] In these studies, rats conditioned with a heat stress sufficient to cause HSP accumulation survived a subsequent and otherwise lethal exposure to intravenous bacterial endotoxin compared to naïve control animals. Subsequent studies in peritoneal macrophages demonstrated that the induction of a stress response as measured by cellular HSP70 accumulation resulted in the decreased transcription and secretion of the inflammatory cytokines tumor necrosis factor-α (TNF-α) and Interleukin-I-β (IL-1β). This effect was not seen with Interleukin-6 (IL-6), and normal cytokine production and secretion resumed with the decay of intracellular HSP70.[5,6] The decrease in cytokine production associated with a decrease in specific mRNA was not caused by changes in either message half-life or translational efficiency, but rather to an effect on cytokine message conscription. The ability of the HSP response to alter transcription may be related to an interference between the primary transcription factor for the heat shock response (HSF1) and important cytokine transcription factor. In an intact animal model, HSF1-deficient mice have been shown to be highly sensitive to endotoxin exposure, and this sensitivity is associated with an augmentation in sound TNF-α.[7] Studies in cells have demonstrated that the activation of HSF1 is associated with a decrease in activation of the important TNF-α transcription factor NF$\kappa\beta$. Interestingly, exposure to LPS results in inactivation of HSF1.[8] It has been proposed that this HSF1 inactivation could be mediated through LPS activation of MAP kinases.[9] HSF1 activation results in the inactivation of important cytokine transcription factors. Conversely, the inflammatory signals such as lipopolysaccharide (LPS) activate these inflammatory transcription factors and block the central HSP transcription factor HSF1.

Unlike LPS, inflammatory cytokines do appear to stimulate HSP production in a variety of systems. In cardiac myocytes, for example, HSP27 and HSP70 accumulation occurs following one exposure.[10] In pancreatic islet cells, HSP70 and HSP90 accumulation occur following one beta exposure.[11] The cellular increase in HSPs, in turn, is associated with a state of tolerance not only to heat stress, but to subsequent cytokine exposure. Cells with an impaired stress response, as measured by cellular HSP70 accumulation, show increased TNF-α-induced cytotoxicity.[12] This altered TNF resistance profile has been demonstrated through HSP70 overexpression studies in cell culture that demonstrate that this overexpression blocks TNF cytotoxicity.[13,14] The overexpression of HSP70 and subsequent TNF resistance do not appear to be related to alterations in cellular TNF-α reception expression but does appear to be associated with impaired TNF-α-induced phospholipase A_2 activation.[15] In a previously described study of IL-1β-induced HSP accumulation in pancreatic islet cells, the cells that accumulated HSP70 and HSP90 in response to IL-1β stimulation were rendered resistant to subsequent IL-1β-induced cytotoxicity. Taken as a whole, these data suggest a feedback model for cytokine exposure and cytotoxicity whereby cells exposed to cytokines generate a stress response, and this stress response is associated with subsequent cytokine resistance. The association of HSP accumulation and decreased inflammatory cytokine production by cells suggests an important role for the stress proteins in the alteration of both inflammatory cytokine production and resistance.

The impact of these cellular alterations has been examined in the intact organism. In one study, a conditioning heat stress sufficient to cause HSP70 accumulation of the liver was associated with a decrease in serum TNF-α but no change in serum IL-6 following intravenous LPS exposure.[16] That the liver is a critical target for heat conditioning was demonstrated in a subsequent study in which liver protein synthesis was blocked during the conditioning heat stress.[17] Despite the fact that organs other than the liver accumulated HSP70 as a result of a heat stress, the subsequent LPS exposure resulted in a marked increase in circulating TNF-α and refractory hypothermia. These animal models are consistent with the cellular data regarding cytokine production and resistance, and further suggest that specific organs such as the liver may be important sites for these effects in the intact organism.

B.   HSPs in the extracellular microenvironment

Given the important association of HSP accumulation with decreased cytokine production and increased cytokine resistance, and the ability of cytokines and LPS to modulate the HSP response, it is perhaps surprising that HSPs seen by immune effector cells in the extracellular environment mediate a vigorous and multifaceted immune system activation. This activation is driven by two different HSP phenomena: (1) the modulation of natural killer cells cytotoxicity by HSPs displayed on the surface of target cells, and (2) the activation of cytotoxic T-lymphocytes by HSP-peptide complexes through a mechanism requiring endogen pristine cells.

1.   HSP surface expression

Early observations in tumor cells[18] demonstrated that human tumor cell lines showed unusual compartmentalization of several HSP families, primarily HSP90 and HSP70. In addition to finding abnormal nuclear localization of HSP72, this same protein was found on the surface of tumor cells. Other studies had demonstrated members of the HSP90 family[19] as well as HSP70[20] on tumor cells. Subsequently, studies in Ewing's sarcoma and osteosarcoma cells demonstrated HSP72 on the cell surface that could be increased by a conditioning heat stress. Moreover, this increase in surface HSP72 correlated with the recognition and killing of these tumor cells by CD3-negative natural killer (NK) cells but not CD3-positive cytotoxic T-lymphocytes. HSP72 could not be demonstrated on a limited number of nonmalignant cell lines.[21,22] Further studies in a human carcinoma cell line demonstrated that the separation of these cells using HSP72 surface expression resulted in a population of cells that were highly sensitive to natural killer cell lysis and a more resistant population of natural killer cell lysis, again correlating with whether these cells were positive or negative HSP72 surface expressors, respectively.[22] The close association between HSP72 surface expression and cytotoxicity mediated by activated NK cells was further demonstrated in studies involving tumor cells that were both negative and positive for surface HSP72 following a conditioning heat stress. Tumor cells that did not express HSP72 following heat showed no difference in NK-mediated cytotoxicity between heat-conditioned and unconditioned cells. In contrast, tumor cells with HSP72 surface expression following a conditioning heat stress showed increased specific NK-mediated cytotoxicity following this conditioning heat stress which could be blocked by the preincubation of target cells with the antibody specific for HSP72.[23] HSP70 surface expression appears not only in tumor cell lines, but also on primary biopsy material from leukemic patients. As seen in cell culture models, the presence of surface HSP70 appears to correlate with killing of the tumor cells by Interleukin-2 stimulated natural killer cells.

Surface expression of HSP70 is not limited to tumor cells alone. Cells infected with the human immune deficiency virus have also been shown to stain positive for surface HSP70; this may reflect the involvement of HSP70 in viral particular release. In other studies, HSP60 has been demonstrated on the surface of stressed but otherwise normal aortic endothelial cells.[24] The surface expression of HSPs such as GP96 appears to have been conserved across species from human liver cells to *Xenopus* lymphocytes.[25,26] The significance of surface expressed HSPs on otherwise normal cells is unknown. The mechanism by which the HSPs, especially members of the HSP70 family, are displayed on the cell surface is unclear because HSPs do not possess transmembrane domains. Studies using chemotherapeutic agents such as alkyl-lysophospholipids[27] and etherlipids[28] demonstrate increased natural killer cell killing and/or increased HSP70 surface expression on tumor cells. Common mechanisms by which both heat and chemotherapeutic agents might increase HSP70 surface expression include possible membrane alteration as well as the generation of abnormal peptide fragments containing transmembrane domains with subsequent transport to the cell membrane by HSP70 family members. Another interesting possibility is that the HSP70 found on the cell surface, which is antigenically similar to cytosolic HSP70, could in fact be a distinct and novel peptide. Previous studies[29] have suggested that antibody recognition of a surface HSC70-like molecule in fact represents a novel protein that is localized to the plasma membrane. To date, these findings have not been extended to the highly heat inducible HSP72 expressed on the cell surface.

2. HSP-peptide complexes

While it has long been recognized that tumor cell lysates could be used in model systems to induce tumor regression, a recent body of data has suggested that the tumor cell lysate fraction most associated with tumor cell regression is that containing HSP70, HSP90, and the endoplasmic reticulum chaperone GP96.[2,3,30] Using a variety of model systems, investigators have demonstrated that these HSP-tumor peptide fractions, either purified from tumor cell lysates or generated *in vitro*, result in tumor cell regression.[31,32] The ability of these HSP-peptide complexes to cause tumor regression is based on both the HSP and the peptide carried by the HSP. In this regard, the stripping of peptides from the HSP using ATP results in a loss of immune response to the tumor. And, while immunization with tumor peptides alone can be effective in causing regression of implanted tumors, the HSP-peptide complexes are many-fold more potent in generating this response. The HSPs appear to bind to a variety of precursor peptides of tumor cells, which may provide the immune system with a wide variety of target molecules.[33]

The mechanism by which HSP-peptide complexes induce tumor regression have been extensively evaluated. Current data support the fact that HSP-peptide complexes require uptake and presentation by antigen-presenting cells for presentation to cytotoxic T-lymphocytes.[34,35] Recent data suggests that in the case of the endoplasmic reticulum chaperone GP96, uptake in antigen presenting cells occurs through the α2-macroglobulin (CD91) receptor.[36] In addition to HSP70, HSP90, and GP96, other HSPs appear important in driving an immune response to tumors. These include HSP65,[37] HSP110, and the glucose regulator protein GRP170.[38]

Regardless of HSP type, when taken up by dendritic cells appeared to generate a maturation signal within the dendritic cell and to subsequently activate an NF-kβ signaling pathway similar to that of bacterial lipopolysaccharide.[39] The ability of these HSPs to stimulate such a signal has been indirectly demonstrated by the ability of severe hyperthermia to dramatically augment TNF-α gene expression in myelomonocytic cell lines.[40] Further, in tumor implantation models, HSP70 overexpression and subsequent release from tumor cells results in dendritic cell maturation with enhanced antigen uptake, T-lymphocyte activation, and release of inflammatory cytokines such as TNF-α, Interleukin-12, and γ-interferon.[41]

Based on these data, a variety of HSP-peptide complexes are preferentially taken up by antigen presuming cells, perhaps through specific receptor pathways like the α2-macroglobulin receptor as described for GP96. Upon uptake, antigen presenting cells undergo maturation inactivation, perhaps through an NF-kβ pathway. The HSP associated peptide is then presented to key cytotoxic lymphocytes which undergo activation with resultant target cell killing.

IV. Examining the HSP-immune modulation dichotomy

The data presented in the previous section describes the important role of activation of the heat shock response in down-regulating the immune response. This correlation is particularly strong with regard to inflammatory cytokine production. The fact that transcriptional activators of the heat shock response appear to be involved in down-regulation of the cytokine pathway activated by LPS reinforces the strong interrelationship between HSP activation and inflammatory cytokine production. Moreover, we have seen that the induction of the heat shock response in target cells, as a consequence of either a stress stimulus or through specific HSP gene overexpression, renders these cells resistant to the cytotoxic effects of the same cytokines down-regulated by HSP gene activation in immune effector cells. Given this potent and relatively global effect of intracellular HSPs in inhibiting inflammatory cytokine production and rendering target cells cytokine resistant, the pro-inflammatory and immune activating effects of HSPs seen in the extracellular microenvironment would seem to be inconsistent. These effects, related either to HSPs expressed on the surface of cells or released from injured or dying cells, reflects an equally global and potent activation of a variety of immune effector cells as the down-regulation of immune system components by intracellular HSPs.

How do we reconcile this apparent paradox regarding the role of HSPs in immune system activation and inflammation? The answer may well lie in the examination of the evolutionary biology of the HSPs, the immune recognition based on the interpretation of danger signals, and the possible role of viruses in molding the immune system's recognition and response patterns.

One way to examine the role of HSPs as immune system activators is to examine the immune response from the standpoint of the danger paradigm of immunity proposed by Matzinger.[42,43] This danger theory of immunity is offered as a way to understand the limitations of the theory of immune function which relies on distinguishing between "self and non-self." The danger theory of immunity as described by Matzinger states that the self vs. non-self paradigm does not adequately address immune tolerance to the normal changes in protein production and release which occur during development. In the case of puberty, for example, new proteins are produced by the organism at a time when the immune system is considered to be mature. The result of these new proteins in the setting of an already committed immune system might well result in a vigorous auto-immune response against these new proteins which would be considered non-self because of their relative stage of development in terms of the immune system. Instead, Matzinger suggests that the immune system is less concerned with distinguishing self from non-self than it is in distinguishing danger from no danger. Danger signals are elaborated as cells undergo an event that results in either necrotic cell death or the production and subsequent release of key danger signals which would normally be released under conditions of necrosis as opposed to apoptosis.

In this regard, apoptotic cell death can be seen as the principal function of all cells and results in the absence of key cellular signals. During apoptosis, stress signals are neither elaborated as a part of the process nor released into the local extracellular microenvironment during cell death.[39] In contrast, cells undergoing ischemia, necrosis, hyperoxia, or cytokine exposure generate a stress response with the stimulation of intracellular HSPs. These intracellular HSPs may play a role in the cell's attempt to preserve itself during the stress. If unsuccessful, the resulting cell death would result in the release of relatively large amounts of these stress proteins into the local extracellular microenvironment. Once released, these stress proteins would be available for uptake and evaluation by local antigen presenting cells.

Viral infection is another way in which HSPs might be released from the cells. As discussed later, many viral infections are associated with cellular HSP accumulation, usually at an early stage of infection. As viral replication continues to the point of cell death, HSPs would be released from these virally infected cells, as was seen during necrotic cell death. Viruses may also cause the appearance of HSPs in the extracellular microenvironment through the interaction of certain HSPs, principally members of the HSP70 family, with viral particles released from the cell. This HSP70–viral envelope association has been demonstrated with a number of viruses, including the human immunodeficiency virus and the measles virus. In addition and as noted above, members of the HSP70 family have been demonstrated on the surface of cells infected with viruses such as the human immunodeficiency virus. Thus, the very sorts of cellular responses that would constitute unintended cell death or infection by viruses are the same stresses that activate the HSP response. Within the cell, this HSP response may serve a protective role for the individual cell. Upon cell death or viral release, however, these same stress response proteins might have an important secondary role for the multicellular organism: namely, activation of the immune response. Additionally, in the case of viruses, the release of viral peptide-HSP complexes might provide important antigen information in addition to the HSP danger signal.

In addition to the production and release of HSPs and other normally intracellular proteins as danger molecules, Matzinger has based the danger theory of immunity on three laws of lymphotics. These laws dictate the way in which lymphocytes deal with peptide signals in the presence or absence of danger molecules such as HSPs. For the immune system activation to occur, these three laws must be operational.

- The first law of lymphotics states that T-lymphocytes under resting conditions require both a peptide signal and co-stimulatory signal for activation. This first peptide signal is the binding of a T-cell receptor to the major histocompatibility (MHC) peptide complex on an antigen presenting cell. Within a finite time period, a co-stimulatory event related to the danger signal must occur. In the absence of a co-stimulatory event, stimulation of the T-lymphocyte by antigen stimulation alone does not result in activation but results in T cell apoptosis.

- The second law of lymphotics states that this co-stimulatory event must be mediated directly by the antigen presenting cell. Under the second law, T-lymphocytes would not undergo activation in the absence of direct interaction with danger molecules. The second law of lymphotics means that local antigen presenting cells control the activation or apoptotic response of T-lymphocytes recognizing the antigen stimulus.
- The third law of lymphotics is that T-lymphocytes, once activated, remain in this activated state for a defined period. If, during this period, the lymphocyte is re-stimulated, the stimulatory event can occur through the MHC-peptide complex alone in the absence of a co-stimulatory signal provided by the antigen presenting cell. In this way, the antigen presenting cell controls T-cell activation through a simplified re-stimulation event requiring no further danger signal elaboration.

These laws of lymphotics mean that antigen presenting cells, through their presentation of peptide and co-stimulatory signals, determine both the activation or removal through apoptosis of T-lymphocytes. T-lymphocytes, once stimulated by both the MHC peptide complex and the co-stimulatory signal, proliferate to become cytotoxic T-cells with defined antigen specificity. These T-cells, when stimulated, can move between active required states and within a defined time period and can undergo reactivation by MHC peptide complex alone. Thus, the antigen presenting cell serves as the key link between initial stimulation through primary and co-stimulatory events.

The fact that T-cell stimulation by antigen presenting cells through the MHC peptide complex alone results in apoptosis without a danger signal provides a straightforward explanation for the concept of tolerance. Using puberty as the example, the elaboration of new proteins by cells would result in antigens presented to a mature immune system by T-cells capable of recognizing these antigens. Because no danger signal is elaborated, the response of these specific T-lymphocytes to these new proteins is T-cell apoptosis. The repeated stimulation of T-cells by these new, but non-threat molecules would result in apoptosis of all T-cells capable of recognizing these antigens, leading to a state of tolerance. One other important facet of this model is that cytotoxic T-lymphocytes themselves induce apoptosis in target cells. As a result, cytotoxic T-lymphocyte-mediated cell lysis through apoptosis does not, of itself, generate stress signals. This lack of danger signal production by target cells allows for the immune response to be contained to the neighborhood of the stress response, rather than becoming a self-perpetuating phenomenon.

The danger theory of immunity does not completely explain immune system function, but rather provides both a simple model of tolerance and a paradigm to address the immune activating effects of extracellular HSPs. Regardless of which paradigm is invoked to explain immune function, the role of the HSPs themselves in modulating many facets of the immune response must be evaluated in an evolutionary context. The HSPs are a system, which appeared early in prokaryote evolution and is involved in both cellular house-keeping functions and in the single cell's adaptation to environmental threats. The HSP response can thus be considered as a single cell's default pathway to adapt to hostile changes in its environment. Evolutionary models demonstrate that within the 70,000 molecular weight heat-shock protein family, approximately 75% of the sequence link of these proteins is conserved across all evolution. In eukaryotes, the 70,000 molecular weight family appears to have diverged into the endoplasmic reticulum-associated, glucose-regulated proteins and the cytosolic and nuclear HSP70 family. This divergence and the conservation of sequences supports the concept that members of the HSP70 family existed before the separation of prokaryotes and eukaryotes.

This primitive, yet highly conserved system is activated not only by environmental stresses, but also by numerous viral infections. The span of viruses associated with the HSP response is as broad as the organisms that elaborate this response and include bacteriophage, baculovirus, and numerous human pathogens. Members of divergent virus families appear to have independently evolved ways to co-opt, engage, or manipulate the heat shock response in the course of infection. For example, HSP72 expression increases viral transcription, plaque size, and cytopathogenicity in measles virus-infected cells. In addition, studies using a measles virus mini-replicon system demonstrate that HSP72 overexpression by transfection or heat shock improves the cDNA yield.[44,45] Morbilliviruses incorporate HSP72 into their nucleocapsid, increase their polymerase activity, and demonstrate a requirement for HSP72 for efficient replication.[46,47]

The ubiquity of the HSP response to viral infection suggests that cells respond to viral stress in the same way that they respond to physical stresses—that is, by the accumulation of HSPs. This response could be that the host cells attempt to adapt to the production of foreign (i.e., viral) proteins or the result of generalized transcriptional activation of the cell by viral conscription factors. More recent data, however, demonstrates just the opposite; HSP gene expression induced early in viral infection is, in fact, a viral scheme essential for replication. Such a viral-mediated scheme could be an effective viral strategy to direct the specific activation of a limited set of heat shock proteins for use in the transport and assembly of viral proteins. This latter hypothesis was recently tested using a unique avian adenovirus (the chicken embryo lethal orphan (CELO) virus) that encodes an anti-apoptotic gene, termed gallus anti mort (Gam-1).[48,49] Gam-1 has as an essential function: the activation of the HSP70 and HSP40 genes with subsequent nuclear relocalization. In this series of studies, Gam-1 was removed from the viral genome and its function, resulting in no viral replication.[50] The function of Gam-1 could be replaced by the overexpression of Gam-1 with the replication-deficient virus, as well as by a heat stress sufficient to induce a heat shock response or by the overexpression of HSP40 with the replication-deficient virus. Numerous viruses, including herpes simplex, cytomegalovirus, measles, and adenovirus, all induce heat shock protein accumulation early in infection. Given that bacterial HSP70 and HSP40 homologues are essential for the replication of bacteriophage lambda[51] the ability of a specific HSP, in this case the HSP70 co-factor HSP40, to replace the function of a viral gene in replication, demonstrates the essential function of at least one heat-shock protein in viral replication.

When viruses utilize the translational machinery of cells to synthesize viral proteins, they may require the production of inducible or sometimes constitutive HSP70 to assist with proper folding, assembly, and membrane transport of viral proteins. Thus, active engagement of cellular HSPs for their chaperone function, and/or as a means to contain the potent peptide displaying activity of HSPs, may be fundamental for many or even all viruses.[52] While much of the data regarding viral protein–HSP interaction focuses on HSP70 and its co-factors such as HSP40, HSPs other than those in the HSP70 class can be involved. Such an example is the involvement of HSP90 in the nucleic acid binding activity of hepatitis B virus (HBV) polymerase.[53] As a strictly *in vitro* phenomenon, highly disparate viruses have been found to reactivate from a latent or at least repressed state as a consequence of activation of the stress response pathway.[54] Our recent demonstration that HSP40 can replace the function of the adenoviral anti-apoptotic factor Gam-1, along with a similar observation for a functional HSP70-like activity in HIV-1,[55] strengthens the growing perception that viruses may be all but obliged to subvert or regulate HSP activities as a central facet of their replication strategies.[56,57] In fact, some viruses, such as the beet yellow virus, even encode HSP homologues to assure the proper folding, transport, and assembly of their proteins.[58] Similarly, the SV40 T antigen functions like the bacterial HSP40 homologue DnaJ.[59,60]

Given this recently demonstrated essential role for certain HSPs in viral replication, an evolutionary model can now be put forth to reconcile the seemingly opposite roles of intracellular and extracellular HSPs immune system suppression and activation, respectively.

Evolutionary genetic models support the notion that the HSP70 family existed before the separation of prokaryotes from eukaryotes. The HSP70 family of protein chaperones, assemblers, and transporters were functioning at an evolutionary time remote from either organized multicellular life or the presence of a cellular immune system. During this early period, viral co-evolution may have resulted in the development of a replication strategy to subvert this cellular machinery involved in protein transport and assembly to aid in viral protein management and subsequent production of virions. Further, data from other viral systems suggest a role for HSP40 and HSP70 homologues in viral gene expression, protein movement, envelope production, and extrusion from the cell. Given the role of bacterial HSP70 and HSP40 homologues in phage replication, viral infection of contemporary prokaryotes appears to function in this manner.

With the advent of multicellular organisms, specialized systems including the immune system developed in the context of these HSP-associated viral replication strategies. Under these conditions, an effective immune counter-strategy to viral replication would include the recognition in the extracellular environment of these otherwise intracellular and normally stress regulated HSPs as a signal for potential viral infection. The natural result of this immune system counter-strategy means that HSPs can be viewed as part of a normal immune response because the members of the HSP70 family, the HSP90 family, and the endoplasmic reticulum chaperone family gp96 can display viral peptides as virally infected cells are disrupted, thus greatly intensifying an immune response against the foreign proteins. The subsequent development of HSPs, as important molecules for immune system activation under a variety of circumstances, and the potential employment of HSP-peptide complexes as agents to induce specific tumor immunotherapy, are consistent with this viral infection paradigm.

V. Conclusions

The role of HSPs in multicellular organisms appears to have expanded from that in the single cell to participate in modulating the immune/inflammatory response. The association of intracellular HSPs with a down-regulation of cytokine production and increased cytokine resistance contrasts with the role of HSPs on the cell surface and as HSP-peptide complexes in immune activation. The fact that numerous viruses infecting both single-cell and multicellular organisms evolved to utilize the HSPs for replication provides a rationale for the immune activating effects of HSPs in the extracellular microenvironment. Understanding the consequences of HSP accumulation within the cell and in the extracellular environment may require investigations in both the cellular biology and evolutionary biology of the adaptive responses of both single-cell and multicellular organisms.

Acknowledgments

The author thanks Brian Hjelle, M.D.; Craig Jensen, M.D.; Karla Melendez; Antonio Panganiban, Ph.D.; and Eric Wallen for their editorial assistance. Research described from the author's laboratory is supported by NIH AR40771, AG14687, and HL61389.

References

1. Welte, M. A. et al., A new method for manipulating transgenes: engineering heat tolerance in a complex, multicellular organism, *Curr. Biol.,* 3, 842, 1993.
2. Skowyer, D., Georgopoulos, C., and Zylicz, M., The *E. coli* dnaK gene product, the HSP70 homolog, can reactivate heat-inactivated RNA polymerase in ATP hydrolysis-dependent manner, *Cell,* 62, 939, 1990.
3. Parsell, D. A. and Lindquist, S., Heat shock proteins and stress tolerance, in Biol Heat Shock Proteins and Molecular Chaperones, CSHL Press, 18, 457, 1994.
4. Ryan, A. J. et al., Acute heat stress protects rats against endotoxin shock, *J. Appl. Physiol.,* 73, 1517, 1992.
5. Ensor, J. E. et al., Differential effects of hyperthermia on macrophage Interleukin-6 and TNFα expression, *Am. J. Physiol.,* 266 (Cell. Biol. 35), C967, 1994.
6. Snyder, Y. M. et al., Transcriptional inhibition of endotoxin-induced monokine synthesis following heat shock in murine peritoneal macrophages, *J. Leukocyte Biol.,* 51, 181, 1992.
7. Xiao, X. et al., HSF1 is required for extra-embryonic development, postnatal growth and protection during inflammatory responses in mice, *EMBO J.,* 18, 5943, 1999.
8. Singh, J. S. et al., Inhibition of tumor necrosis factor-α transcription in macrophages exposed to febrile range temperature: a possible role for heat shock factor-1, *J. Biol. Chem.,* 275, 9841, 2000.
9. Hasday, J. D. and Singh, I. S., Fever and the heat shock response: distinct, partially overlapping processes, *Cell Stress Chaperones,* 5, 471, 2000.
10. Löw-Friedrich, I. et al., Cytokines induce stress protein formation in cultured cardiac myocytes, *Basic Res. Cardiol.,* 87, 12, 1992.
11. Helqvist, S. et al., Heat shock protein induction in rat pancreatic islets by recombinant human interleukin 1b, *Diabetologia,* 34, 150, 1991.
12. Müller, E. et al., Interaction between tumor necrosis factor-α and HSP70 in human leukemia cells, *Leukocyte Res.,* 17, 523, 1993.
13. Jäätelä, M. et al., Major heat shock protein HSP70 protects tumor cells from tumor necrosis factor cytotoxicity, *EMBO J.,* 11, 3507, 1992.
14. Jäätelä, M. and Wissing, D., HSPs protect cells from monocyte cytotoxicity: possible mechanism of self-protection, *J. Exp. Med.,* 177, 231, 1993.
15. Jäätelä, M., Overexpression of major heat shock protein HSP70 inhibits tumor necrosis factor-induced activation of phospholipase A_2, *Immunology,* 151, 4286, 1993.
16. Kluger, M. J. et al., Effect of heat stress on LPS-induced fever and tumor necrosis factor, *Am. J. Physiol.,* 273, R858, 1997.
17. Dokladny, K. et al., Effect of heat stress on LPS-induced febrile response in D-galactosamine-sensitized rats, *Am. J. Physiol.,* 280, R338, 2001.
18. Ferrarini, M. et al., Unusual expression and localization of heat-shock proteins in human tumor cells, *Int. J. Cancer,* 51, 613, 1992.
19. Ferrarini, M. et al., Distinct pattern of HSP72 and monomeric laminin receptor expression in human lung cancers infiltrated by g/d T lymphocytes, *Int. J. Cancer,* 57, 486, 1994.
20. Konno, A. et al., Heat- or stress-inducible transformation-associated cell surface antigen on the activated H-ras oncogene-transfected rat fibroblast, *Cancer Res.,* 49, 6578, 1989.

21. Multhoff, G. et al., CD3⁻ large granular lymphocytes recognize a heat-inducible immunogenic determinant associated with the 72-kD heat shock protein on human sarcoma cells, *Blood*, 86, 1374, 1995.
22. Multhoff, G. et al., Heat shock protein 72 on tumor cells, *J. Immunol.*, 158, 4341, 1997.
23. Roigas, J. et al., Heat shock protein (HSP72) surface expression enhances the lysis of a human renal cell carcinoma by IL-2 stimulated NK cells, in *Gene Therapy of Cancer*, Walden et al., Eds., Plenum Press, New York, 1998, 225–229.
24. Xu, Q. et al., Surface staining and cytotoxic activity of heat-shock protein 60 antibody in stressed aortic endothelial cells, *Circ. Res.*, 75, 1078, 1994.
25. Robert, J., Menoret, A., and Cohen, N., Cell surface expression of the endoplasmic reticular heat shock protein gp96 is phylogenetically conserved, *J. Immunol.*, 163, 4133, 1999.
26. Altmeyer, A. et al., Tumor-specific cell surface expression of the KDEL containing endoplasmic reticular heat shock protein GP96, *Int. J. Cancer*, 69, 340, 1996.
27. Botzler, C. et al., Noncytotoxic alkyl-lysophospholipid treatment increases sensitivity of leukemic K562 cells to lysis by natural killer (NK) cells, *Int. J. Cancer*, 65, 633, 1996.
28. Botzler, C. et al., Synergistic effects of heat and ET-18-OCH3 on membrane expression of HSP70 and lysis of leukemic K562 cells, *Exp. Hematol.*, 27, 470, 1999.
29. Hirai, I. et al., Localization of pNT22 70 kDa heat shock cognate-like protein in the plasma membrane, *Cell Struct. Funct.*, 23, 153, 1998.
30. Li, Z. and Srivastava, P. K., Tumor rejection antigen GP96/GRP94 is an ATPase: implications for protein folding and antigen presentation, *EMBO J.*, 12, 3143, 1993.
31. Blachere, N. E., et al., Heat shock protein vaccines against cancer, *J. Immunother.*, 14, 352, 1993.
32. Ciupitu, A.-M. T. et al., Immunization with a lymphocytic choriomeningitis virus peptide mixed with heat shock protein 70 results in protective antiviral immunity and specific cytotoxic T lymphocytes, *J. Exp. Med.*, 187, 685, 1998.
33. Ishii, T. et al., Isolation of MHC class I-restricted tumor antigen peptide and its precursors associated with heat shock proteins HSP70, HSP90, and GP961, *J. Immunol.*, 162, 1303, 1999.
34. Udono, H., Levey, D. L., and Srivastava, P. K., Cellular requirements for tumor-specific immunity elicited by heat shock proteins: tumor rejection antigen GP96 primes CD8+ T cells *in vivo*, *Proc. Natl. Acad. Sci. U.S.A.*, 91, 3077, 1994.
35. Srivastava, P. K. et al., Heat shock proteins transfer peptides during antigen processing and CTL priming, *Immunogenetics*, 39, 93, 1994.
36. Binder, R. J. et al., Saturation, competition, and specificity in interaction of heat shock proteins (HSP) GP96, HSP90, and HSP70 with CD11b+ cells, *J. Immunol.*, 165, 2582, 2000.
37. Lukas, K. V. et al., *In vivo* gene therapy of malignant tumours with heat shock protein-65 gene, *Gene Therapy*, 4, 346, 1997.
38. Wang, X.-Y. et al., Characterization of heat shock protein 110 and glucose-regulated protein 170 as cancer vaccines and the effect of fever-range hyperthermia on vaccine activity, *J. Immunol.*, 165, 490, 2001.

39. Basu, S. et al., Necrotic but not apoptotic cell death releases heat shock proteins, which deliver a partial maturation signal to dendritic cells and activate the NF-κB pathway, *Int. Immunol.*, 12, 1539, 2000.

40. Iwamoto, G. K., Ainsworth, A. M., and Moseley, P. L., Hyperthermia enhances cytomegalovirus regulation of HIV-1 and TNF-α gene expression, *Am. J. Physiol.*, 277 (Lung Cell. Mol. Physiol.) 21, L1051, 1999.

41. Todryk, S. et al., Heat shock protein 70 induced during tumor cell killing induces Th1 cytokines and targets immature dendritic cell precursors to enhance antigen uptake, *J. Immunol.*, 163, 1398, 1999.

42. Matzinger, P., An innate sense of danger, *Semin. Immunol.*, 10, 399, 1998.

43. Matzinger, P. and Fuchs, E., Beyond "self" and "non-self" immunity is a conversation, not a war, *J. NIH Res.*, 8, 35, 1996.

44. Parks, C. L. et al., Enhanced measles virus cDNA rescue and gene expression after heat shock, *J. Virol.*, 73, 3560. 1999.

45. Vasconcelos, D. Y., Cai, X. H., and Oglesbee, M. J., Constitutive overexpression of the major inducible 70 kDa heat shock protein mediates large plaque formation by measles virus, *J. Gen. Virol.*, 79(Pt. 9), 2239, 1998.

46. Oglesbee, M. J. et al., Enhanced production of morbillivirus gene-specific RNAs following induction of the cellular stress response in stable persistent infection, *Virology*, 192, 556, 1993.

47. Oglesbee, M. J. et al., The highly inducible member of the 70 kDa family of heat shock proteins increases canine distemper virus polymerase activity, *J. Gen. Virol.*, 77(Pt. 9), 2125, 1996.

48. Chiocca, S., Baker, A., and Cotton, M., Identification of a novel antiapoptotic protein, Gam-1, encoded by the CELO adenovirus, *J. Virol.*, 71, 3168, 1997.

49. Michou, A.-I. et al., Mutational analysis of the avian adenovirus CELO, which provides a basis for gene delivery vectors, *J. Virol.*, 73, 1399, 1999.

50. Glotzer, J. B. et al., Activation of heat shock response by an adenovirus protein is essential for virus replication, *Nature*, 407, 207, 2000.

51. Polissi, A., Goffin, L., and Georgopoulos, C., The *Escherichia coli* heat shock response and bacteriophage lambda development, *FEMS Microbiol. Rev.*, 17, 159, 1995.

52. Saphire, A. C. et al., Nuclear import of adenovirus DNA *in vitro* involves the nuclear protein import pathway and HSC70, *J. Biol. Chem.*, 275, 4298. 2000.

53. Hu, J. and Anselmo, D., *In vitro* reconstitution of a functional duck hepatitis B virus reverse transcriptase: posttranslational activation by HSP90 [In Process Citation], *J. Virol.*, 74, 11447, 2000.

54. Moriya, A. et al., Heat shock-induced reactivation of herpes simplex virus type 1 in latently infected mouse trigeminal ganglion cells in dissociated culture, *Arch. Virol.*, 135, 419, 1994.

55. Agostini, I. et al., Heat-shock protein 70 can replace viral protein R of HIV-1 during nuclear import of the viral preintegration complex, *Exp. Cell Res.*, 259, 398, 2000.

56. Stanley, S. K. et al., Heat shock induction of HIV production from chronically infected promonocytic and T cell lines, *J. Immunol.*, 145, 1120, 1990.

57. Andrews, J. M. et al., The cellular stress response enhances human T-cell lymphotropic virus type 1 basal gene expression through the core promoter region of the long terminal repeat, *J. Virol.*, 71, 741, 1997.

58. Napuli, A. J., Falk, B. W., and Dolja, V. V., Interaction between HSP70 homolog and filamentous virions of the beet yellows virus, *Virology*, 274, 232, 2000.

59. Campbell, K. S., et al., DNAJ/HSP40 chaperone domain of SV40 large T antigen promotes efficient viral DNA replication, *Genes Dev.*, 11, 1098, 1997.
60. Kelley, W. L. and Georgopoulos, C., The T/t common exon of simian virus 40, JC, and BK polyomavirus T antigens can functionally replace the J-domain of the *Escherichia coli* DnaJ molecular chaperone, *Proc. Natl. Acad. Sci., U.S.A.*, 94, 3679, 1997.

chapter eleven

Stress proteins and applied exercise physiology

Jürgen M. Steinacker and Yuefei Liu

Contents

I. Introduction

Stress proteins represent one of the general molecular protective mechanisms that enable cells and whole organisms to survive stress. Several stress proteins have been identified which have different functions within cells and tissues and they exhibit early, intermediate, and late patterns of response. Because cellular and tissue stresses are complex processes that involve several cellular stress response mechanisms, the study of their mechanisms is difficult. Physical exercise challenges an organism in an intricate manner and is an important and natural stress for humans. The induction of stress proteins by exercise is associated not only with the temperature-related effects of exercise but with other physiological processes, such as contractile stress, energy metabolism, induction of metabolic and stress-related messengers, and perfusion and oxygen delivery disturbances.[1]

Nearly two decades ago, an attempt was made to determine the effect of exercise on the induction of heat shock proteins (HSPs).[2] Consequently, it has been demonstrated that physical exercise or chronic electrical stimulation induces the expression of HSPs in skeletal muscles[3–6] and that exercise-induced HSP70 production can take place in blood cells,[7–9] liver, heart, and skeletal muscle.[10–13] It has been shown that mitochondrial HSPs, including HSP60 and GRP75, can also be induced in skeletal muscles by exercise, and that the expression of individual mitochondrial HSPs is independently regulated and uncoordinated.[14] Furthermore, HSPs may be influenced by several hormones like estrogen and cytokines like TNF-α and may themselves act as cytokines.

All these functions of HSPs make them relevant and important for the organismic adaptation to acute and chronic stress during training. This review deals with the physiological role of HSPs in the acute exercise and training response in humans.

II. Induction of heat shock proteins in physical exercise

The induction of the HSPs in the muscle during exercise is closely related to contractile activity, the induction of myogenic regulatory factors, and to myofibrillar protein synthesis and degradation.[15–19] Several classic factors speculated to result in the stress response are temperature variation, contraction-related stress, energy metabolism, hormones and cytokines, and perfusion-related stress.

A.   Temperature

It is known that changes in body temperature, which occur during physical exercise, can lead to an HSP response. The ability of stress proteins to confer thermotolerance in both cultured cells and in animals is well-documented,[20,21] and exercise-induced hyperthermia is an important inducer for HSPs.[22] The level of HSPs in HSP-expressing cells may reflect previous hyperthermia- and exercise-related stress. It has been shown that the HSP response on leukocytes was enhanced by additional thermal stress and an augmented HSP mRNA response after a previous hyperthermic exercise has been indicated.[9] However, it has also been shown that there is a temperature-independent effect

of exercise on accummulation of HSP-70 in rat liver and that the temperature effect seems to attenuate with aging.[23] Also, the production of other HSPs such as small HSPs during contractile activity can be temperature independent.[16]

In humans, higher increases in inducible HSP70 in skin fibroblasts of heat-adapated Turks, compared to Russians living in a moderate climate, has been reported, and the increase was related to maintenance of a normal rate of protein synthesis under elevated temperatures.[24]

B. Contraction-related stress

Calcium- and stretch-mediated transcriptional pathways during exercise result in increased protein synthesis and degradation due to the exercise-induced transition of protein isoforms like myosin heavy chains and may therefore change the free to bound HSP ratio.[25] Free calcium levels also increase the effects of protein kinase C-dependent pathways on HSP accumulation.[26] HSF activation has been demonstrated in exercising heart muscle[17] and skeletal muscle,[27] and it may be speculated that this is the dominant pathway for HSP regulation during low to moderate metabolic activity such as endurance training.

The maintenance of structural proteins may be another important mechanism for HSP-mediated stress tolerance, both for intracellular and membranous microfilaments. HSP27, a protein from the small HSP class homologous with αB-crystallin, prevents actin microfilament disruption under stress.[28,29] αB-Crystallin and HSP27 are associated with microfilaments,[30] and αB-crystallin is localized to the Z-lines of slow muscle fibers, increases with contractile activity, and decreases with inactivity.[16,31] Overexpression of αB-crystallin and HSP27 protects cardiomyocytes from ischemic stress.[30]

C. Energy metabolism

Exercise leads to disturbances in cellular oxidative metabolism and to ATP and glycogen depletion. Furthermore, intense exercise is associated with lactate accumulation and acidosis, which cause severe disturbances in cellular homeostasis.

In myogenic cultered cells, ATP depletion to 30% of control levels sufficient for transcriptional activation of HSP, whereas a decrease of STP approximately 50% of control levels did not provoke an increase in HS activation.[32] In these experiments, a reduced pH (6.7), in the face of maintain ATP levels, did not activate HSF.[25] Therefore, ATP depletion, independe of acidosis, may be an important metabolic pathway for HSF activatic during exercise. When ATP content is reduced to a critical level, HSPs rem complexed to unfolded proteins and cannot be recycled. As a consequei the pool of free HSP decreases, HSF forms trimers and HSP gene activat occurs.[25,32,33] Similiar regulation mechanisms have also been demonstr for the small HSPs.[16]

In the whole organism, the metabolic effects of ATP and glycogen deple-tion, acidosis, and lactate accumulation cannot be clearly distinguished. However, studies discussed later clearly point to an intensity-related effect of HSP induction.

Prolonged exercise, like a marathon, leads to induction of oxidative free radicals and an increase in DNA damage and small antioxidative HSPs like HO-1, HSP27, and in HSP70 expression on circulating leukocytes.[34] The res-ponse is attenuated with training, indicating some kind of stress adapation; however, the relevance of this effect for other organs is still under discussion. Recently, it has been demonstrated that prolonged, nondamaging exercise raised the antioxidant enzyme superoxide dismutase (SOD) and HSP60 and 70 in human skeletal muscle and that superoxide dismutase and HSP60 peaked on the second and third day after the exercise, and that HSP70 peaked only at the sixth day of the recovery.[35] Modulation of the oxidative stress by tocopherol supplementation in humans did not affect HO-1 expression in leukocytes before or after exercise,[36] and vitamin E deprivation in rats led to lower HSP70 expression in rat muscle after training.[37] Therefore, it seems likely that the oxidative stress of training and exercise has some effect on HSP expression but is not sufficient to induce HSP70 protein to a great extent.

D.   Metabolic and stress-related messengers

To date, the most relevant effect of hormones on the HSP70 response is that for estradiol. Female rats have significantly lower post-exercise HSP levels than males and, in acutely exercised male rats, estrogen supplementation attenuates the HSP70 response.[38] It has also been demonstrated that the expression of the constitutive form of HSC70 (HSP73) is gender dependent.[39] In the latter study, HSP70 was expressed in a gender-independent manner, but the anabolic steroid nadronolonedecanoate, which is an estrene deriva-tive, increased HSP70 in fast twitch fibers coupled with the anabolic effect of the steroid.[39] HSP90 is involved in the steroid receptor expression and Ah receptor signaling pathways, which may involve regulation of the estrogen and cortisol effects,[40,41] important for the training response. Furthermore, HSP90 is involved in the receptor/mitogen-activated protein kinase (MAPK) signaling pathway[42] and in the endothelial NO synthase,[43] mechanisms that may be relevant for the acute exercise response and training adaptation. Unfortunately, at present, there are no studies of this type using human subjects.

In cell culture, it has been demonstrated that several isoforms of the HSP27 family are induced by inflammatory ctyokines such as TNF-α.[44] It has also been demonstrated that the increase in TNF-α and IL-8 parallels the changes in several HSPs expressed on circulating leukocytes.[45] The possibil-ity that HSPs may themselves act as cytokines in the exercise response is raised by the provocative finding that exogeneous HSP70 stimulated TNF-α, IL-1β, and IL-8 in human monocytes and that both CD14-dependent and

calcium-dependent pathways were activated by HSP70, the former resulting in an increase in TNF-α, IL-6, and IL-8, the latter in TNF-α only.[46]

E.   Perfusion and oxygen delivery

The peripheral skeletal muscles may be subjected to an exercise- or contraction-related hypoperfusion. The muscle contraction itself leads to pulsatile compression of arteries and arterial inflow. This effect and limited perfusion capacity in case of high demand whole-body exercise, when muscle blood flow demand exceeds cardiac output, will lead to hypoperfusion, even in healthy subjects. These effects are aggravated in patients when blood flow is critically restricted by local stenosis or by general hypoperfusion related to low cardiac output or shock, and may lead to local ischemia. Ischemia, hypoperfusion, and hypoxia, which cause ATP depletion, thereby resulting in HSF-activation as described above.[1,18,32,47]

III. The response of heat shock proteins in physical exercise

Considering that the factors inducing heat shock proteins during physical exercise are quite numerous and varied, one would not be very surprised by the results derived from different studies, indicating various responses of heat shock proteins to different types of exercise. The exercise-induced HSP responses may be determined by exercise duration, intensity, and pattern, with differences among species, organs, tissues, and cells in terms of the HSP response.

A.   Acute exercise response

Hammond et al.[2] were the first to attempt to demonstrate that the HSP response could be induced by exercise. Thereafter, several studies demonstrated that the HSP response occurred at both the mRNA and protein level.[3,17]

The HSP response to a short bout of exercise is well-documented.[1] It has been shown that the physiological changes accompanying treadmill running can lead to HSP induction in different tissues.[3,10,17,48] Salo et al.[10] investigated the HSP response by running rats to exhaustion and found significant increase of various HSPs immediately after exercise. Skidmore et al.[48] demonstrated in rats that HSP70 increased in the soleus, gastrocnemius, and the left ventricle, but not in extensor digitorum longus 30 minutes after a 60-minute bout of exercise, and that the accumulation of HSP70 could be independent of body temperature.

In contrast, fewer studies have investigated the HSP response in humans, although there is increasing interest in such studies. Ryan et al.[7] observed that leukocytes which had undergone prior exercise stress, showed a decreased synthesis of HSP72 upon thermal stress. HSP70 (HSP72) is the major inducible

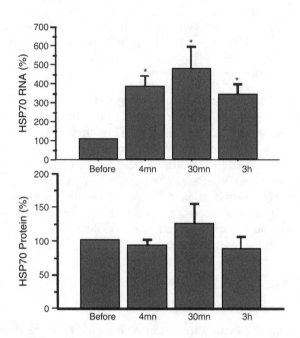

Figure 11.1 Increase in HSP70 mRNA (upper panel) but no change in HSP70 protein (lower panel) in vastus medialis of after a 30-minute run. Mean values and SE. (Adapted from Puntschart, A. et al., *Acta Physiol. Scand.*, 157, 411, 1996.)

isoform of the HSP70 family, and several studies have reported that HSP70 can be induced by exercise in a variety of tissues. Fehrenbach et al.[45] have recently reported that in leukocytes, HSP27 and HSP70 transcripts and protein increased immediately after acute exertion and remained elevated 24 hours after exercise.

Puntschart et al.[5] reported an HSP70 response to acute exercise in human skeletal muscle, as evidenced by an increase in HSP70 mRNA (Figure 11.1). However, they could not demonstrate a significant increase in HSP70 up to 3 hours after exercise (Figure 11.1). Therefore, it remained unclear whether HSP70 in human skeletal muscle could be induced by exercise at all until Liu et al. (1999) reported an HSP70 response in well-trained rowers during a 4-week training period (Figure 11.2). There are several recent studies reporting that HSPs increased in response to acute exercise in humans.[11,35,49] In a study by Thompson et al., eight volunteers (four males and four females, respectively) performed an eccentric exercise with two sets of 25 repetitions and muscle biopsy samples were taken from the exercised biceps brachii 48 hours post-exercise, and the nonexercised biceps served as control. In this study, a dramatic increase of HSC/HSP70 could be shown 48 hours after exercise.[11] The investigators also examined the HSP27 changes in this study, and a clear augmentation of HSP27 over the study period was evident, although the extent of the increase was not as great as that of HSC/HSP70.

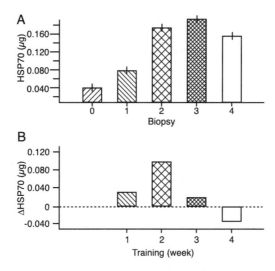

Figure 11.2   HSP70 changes during 4 weeks of training in rowers. (A): compared with before training, HSP70 increased after training began (P = 0.01), reached maximum levels in the third phase, and decreased in the fourth phase. Values are means ±SE. (B): differences (Δ) between two consecutive biopsies, serving as an indirect indicator of HSP70 production, show that maximum HSP70 increase happened at the end of the second training week. Indicated are the training phases: RT: resistance training, INT: intensive training, TAP: tapering, JWC: world championships. (Modified from Liu, Y. et al., *J. Appl. Physiol.*, 86, 101, 1999. With permission.)

Simultaneously, Khassaf et al.[35] reported significant increases in HSP60 and HSP70 after nondamaging one-legged exercise. They observed that HSP60 showed an increase on the second day and peaked on the third day of recovery from exercise, while HSP70 did not peak until the sixth day. The response of HSP60 and HSP70 was also different with regard to their extent of increase.

The above studies suggest that in human skeletal muscle, the HSP response to acute exercise may be different from that found in other animals in which the tissue-specific response to stress[50–52] and exercise[10] is well-documented. In humans, the HSP response to exercise may also differ with different tissues. Fehrenbach et al.[45] reported that HSP70 expression increased in leukocytes immediately after acute exercise and remained elevated for 3 hours, whereas Puntschart[5] did not get an augmented HSP70 level in skeletal muscle although HSP70 mRNA increased significantly. Interestingly, HSP response may be influenced by estrogen[38] and, therefore, the HSP response to acute exercise might be different between males and females.

The variety of HSP response to acute exercise might be, at least partly, a result of the different time course or kinetics of the HSP response. In the study of Puntschart et al.,[5] the observation that HSP70 mRNA but not HSP70 protein increased within 3 hours after exercise could indicate that a single bout of exercise is not strong enough to induce HSP70 in skeletal muscle, or that the period studied post-exercise was too short given the unknown kinetics of the HSP70 response. We have therefore conducted a study to investigate whether there is an HSP70 response in human skeletal muscle to exercise at all. For the first time, we were able to demonstrate that HSP70 increased in human skeletal muscle during physical exercise after 7 days. Thereafter, several studies have reported that a single bout of exercise can induce a HSP70 response in human skeletal muscle.[11,35] This suggests that the results of Puntschart et al.[5] may be best explained by the time course of HSP response. Indeed, Khassaf et al.[35] showed that in response to a non-damaging submaximal exercise, HSP70 increased significantly on day 2 after exercise and peaked on day 6. Thompson et al.[11] demonstrated a significant HSP response in human skeletal muscle 2 days afer exercise. In comparison with animal studies, HSP response or accumulation in human skeletal muscle seems to occur with a somewhat greater time delay. As yet, there is no study reporting HSP induction at the protein level in human skeletal muscle within hours of acute exercise.

Finally, there is a discrepancy between HSP mRNA and protein in terms of the acute exercise response. HSP mRNA usually increases immediately after exercise and remains elevated for a few hours after exercise, whereas an increase of HSP protein has been only observed days after exercise, and the change of HSP mRNA content seems to be more dramatic than that of HSP protein. This is not very surprising because transcription occurs prior to translation. The physiological significance of this discrepancy, due to a time delay of translation, is not clear but it points to a translational control mechanism. Considering that HSPs play a charperone role in maintaining cellular homeostasis, it seems likely—at least for HSP70—that the HSP response to facilitate new protein synthesis, which would happen somewhat later, is more relevant than the involvement in the degradation of denatured proteins being related to the acute phase-response of exercise. Consequently, the HSP response to acute exercise may have a physiological role in skeletal muscle fiber transition, which is worth investigating further.

B.   Heat shock protein response to training

Exercise training can induce a variety of physiological changes, and these changes include the HSP induction factors stated above. Because physical training itself is very complex, varying in duration, volume, intensity, and exercise pattern, the HSP response to physical training exhibits a great variety. Considering the time course of HSP response, it is neither easy nor simple to exactly characterize HSP response to physical training. Until now, little was known regarding HSP response during and after prolonged training. Samelman[53] reported an increase in basal protein levels of HSP72/73 and HSP60 in left and right ventricles and soleus, which remained unchanged in lateral gastrocnemius after endurance training. Powers et al.[54] trained rats for 10 weeks and found that the improvement in the myocardial responses to ischemia-reperfusion was accompanied by a significant increase in HSP72. Furthermore, small HSPs are up-regulated with chronic contractile activity.[15,16]

The responses of HSP to training have been relatively less studied in humans. Fehrenbach et al.[45] investigated HSP response in peripheral leukocytes to training and found that in comparison with the untrained subjects, basal HSP level was lower and HSP response to exercise was higher in the endurance-trained subjects, suggesting an adaptation of HSP response to training and the influence of physical training status on the HSP response. Skeletal muscle HSP response to exercise may be of physiological and pathophysiological significance.[18] It is well-accepted that skeletal muscle HSP can be induced by exercise in animals; however, studies trying to find an HSP response to exercise in human skeletal muscle failed to demonstrate any induction[5] (Figure 11.1). We have conducted the first study able to demonstrate an increase of HSP70 in human skeletal muscle.[6] In this study, we examined ten well-trained rowers who performed a four-week training in preparation for the world championship for rowing. The training program was composed of different phases with different exercise amounts; and at the end of each training phase (1 week each phase), a muscle biopsy was obtained from m. vastus lateralis on the dominant side. HSP70 (inducible form) was quantitatively determined by Western blot with reference to a series of known HSP70 standards. The results showed that HSP70 increased significantly during the entire training period, and HSP70 level changed with different training phases (Figure 11.2). It has been shown that with an increased amount of exercise, net HSP70 production (between two consecutive biopsies) increased, and vice versa. However, in this study it could not be determined whether the relation between exercise amount and HSP response was more dependent upon exercise intensity or exercise volume. Therefore, another study was conducted to deal with this issue.[13] In this study, two groups of athletes underwent different training programs in terms of exercise volume and intensity, and the training program was divided into three phases. In one group, subjects performed higher intensity training in the first phase, with the other group performing such training in the second phase. The exercise volume was increased for both groups in the second phase. The results showed that HSP70 increased significantly when exercise volume and intensity increased; however, the HSP70 level decreased significantly when exercise volume increased but intensity decreased (Figure 11.3). Therefore, it can be concluded that HSP70 response in human skeletal muscle to training is most dependent on exercise intensity. Since in this study, in addition to the differences regarding exercise volume and intensity, different exercise forms (e.g., high-intensity training and endurance training at lower intensity), were employed, the question yet to be answered is whether different exercise forms have different impacts in terms of HSP70 response in human skeletal muscle to training. To address this issue, a further study has been conducted in which six well-trained rowers underwent a training program divided into two phases, followed by one week of regeneration each. The first training phase was mainly high-intensity training with a blood

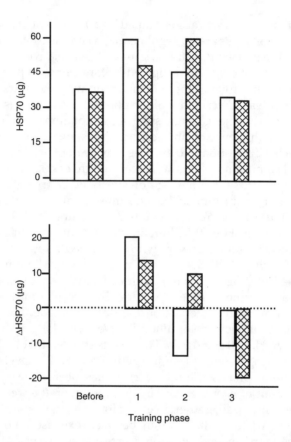

Figure 11.3 HSP70 response to training in vastus medialis of rowers (means). Group A, open bars; group B, hatched bars. Both groups had an identical training volume but differed in exercise intensity in phases 1 and 2. In group A, training intensity was higher in phase 1; in group B, in phase II. Upper panel: Muscle HSP70 content increased in both groups in training phases 1 and 2 (p < 0.05) and then decreased in phase 3. There was no significant HSP70 difference before training and after phase 3 in either group. Lower panel: ΔHSP70 represents the difference in muscular HSP content between two successive biopsies. The increase of HSP70 in group A in phase 1 was higher than in group B (p < 0.05), while ΔHSP70 in phase 2 was higher in group B than in group A (p < 0.05), thereby suggesting the importance of exercise intensity in HSP70 synthesis. (Modified from Liu, Y. et al., *Int. J. Sports Med.*, 21, 351, 2000. With permission.)

lactate concentration above 6 m*M*, and the second phase was mainly endurance training at a lower intensity (blood lactate concentration between 2 and 4 *M*). The exercise volumes of both training phases were matched through low-intensity training and nonspecific training such as gymnastics. The muscle samples were collected from m. vastus lateralis before and at the end of each training phase and regeneration break to determine HSP70 content and

HSP70 mRNA level. The results showed that HSP70 increased significantly after the high-intensity training and decreased during the regeneration period; there was no significant change of HSP70 level over the period of endurance training. HSP70 mRNA showed a significant increase after high-intensity training and decreased gradually thereafter (unpublished data). It was concluded that there are different HSP70 responses to training in human skeletal muscle in terms of high- or low-intensity training, and that differences in the time courses between HSP70 protein and HSP70 mRNA existed, suggesting a potential role of posttranscriptional regulation in the HSP70 response.

Until now there were no data illustrating the relationship between HSP response and performance, nor any data on the biochemical role of HSP in training. It is generally accepted that HSP plays an important role in energy metabolism and cellular repair, which could certainly have important impacts on training as well as physical performance. Training can result in a variety of physiological changes in the skeletal muscle, including muscle fiber transition. There are studies reporting that HSP response is specifically related to muscle fiber types; for example, that HSP70 and small HSPs are indicative of oxidative fibers.[4,14–16] Further study to deal with such issues, especially in human skeletal muscle, would be useful.

C. Exercise-induced hypoperfusion/ischemia and the heat shock protein response

Ischemia can cause a series of cellular changes, including increased intracellular calcium, altered osmotic control, membrane damage, free radical production, decreased intracellular pH, depressed ATP levels, oxygen depletion, and decreased intracellular glucose levels.[55] All these changes have been reported to be able to induce the cellular HSP response.[56]

There are a lot of studies reporting that ischemia can induce HSPs in a variety of tissues, such as myocardium,[57–59] liver,[60] kidney,[61,62] and skeletal muscle.[18,63–65] Investigating the HSP response to ischemia-reperfusion, in skeletal muscle, may be of physiological as well as pathophysiological significance for applied physiology because ischemia-reperfusion during exercise/training may occur. In the contraction phase, muscle may be subjected to ischemia and in the relaxation phase the muscle will be reperfused; therefore, ischemia-reperfusion may have an impact on muscle function. Unfortunately, to date, there has been no report on the HSP response of human skeletal muscle to this phenomenon. Recently, we investigated HSP70 expression in the skeletal muscle of patients with peripheral arterial occlusive disease (PAOD) and found that muscle HSP70 increased significantly in these patients in comparison with that of age-matched non-PAOD controls. Moreover, HSP70 expression differed with the severity of the disease, implying that HSP70 expression is related to the degree of ischemia (unpublished data). HSP70 mRNA showed similar results, indicating the role of HSP70 gene regulation in the ischemic response.

IV. Physiological significance of heat shock proteins in physical exercise

As so-called "molecular charperones," HSPs have been found to be involved in a variety of cellular processes that maintain cellular homeostasis with regard to biochemical metabolism, cellular repair, and defense against stress (for a detailed review, see Reference 56). Hence, HSPs are undoubtedly important in the response to physical exercise and training.

A. Heat shock proteins and exercise-related stress and adaptation

Because HSP can be induced by physical exercise, presumably the HSP response may serve as an indicator for exercise-related stress. Exercise can impact muscle morphology (e.g., inducing muscle hypertrophy) and inactivation or detraining may cause muscle atrophy. There are reports that HSPs play a role in the process of muscle atrophy,[66,67] and HSP70 increased under microgravity, which can lead to muscle atrophy.[68] Aging is another condition issue leading to muscle atrophy. Maiello[69] reported an increase of basal HSP70 in rat kidneys with age. However, HSP70 accumulation in tissues of heat-stressed rats seems to be blunted with advancing age,[23,70,71] and the blunted heat shock protein response to heating observed with aging is probably not a result of the inability to produce inducible HSP72 because older rats revealed a robust response to exertional hyperthermia.[22] These findings are inconclusive, however, as Locke et al.[72] recently demonstrated a normal response of HSF and HSP in skeletal muscle in aged rats in response to heat stress.

Muscle hypertrophy is an important physiological change in terms of muscle adaptation induced by exercise. An increase of HSP72 has been found in the hypertrophied muscle (laboratory model of hypertrophy; for example, compensatory hypertrophy and stretch hypertrophy), but interestingly, such an alteration of HSP72 was not found in the naturally work-induced hypertrophy.[73] Because the postmitotic skeletal muscle has lost its ability to undergo DNA replication, hypertrophy in such muscle means an increase in cell size without cell division. A possible mechanism by which these terminally differentiated cells hypertrophy may involve the induction of cellular protooncogenes and HSP genes, and then a reinduction of the genes normally expressed only in perinatal life, such as fetal isoforms of contractile proteins.[74]

Because studies show that HSP70 can be induced in human skeletal muscle by physical exercise and training and that the HSP70 response is related to exercise intensity, HSPs can provide information about the cellular changes induced by exercise and therefore might serve as an indicator for the exercise intensity in training, which might be of great interest to monitor training and overtraining. Overtraining is defined as an imbalance between training load and regeneration,[75,76] and the resulting processes may theoretically lead to an HSP response,[75] although there are no reports on this subject.

The HSP response to exercise may also protect the cell against stress so that cellular functions can be preserved. Therefore, it would be reasonable that there might be a relationship between the HSP response and physical performance although this relationship is complex. Skeletal muscles of transgenic mice with up-regulated HSP70 displayed no change in fatigue resistance, but did show an increased sensitivity to caffeine[77] Further studies using such models would be of interest.

Physical exercise and training can cause cellular damage and, as in other models, can induce HSP response. Thus, there may be a relationship between muscle damage and HSP induction.[6,18] We have observed at the beginning of resistance training an exercise-induced muscle damage indicated by elevated serum creatine kinase (CK) levels; during training, HSP70 in the skeletal muscle increased, and comcomittent with the increase of muscle HSP70 serum, CK decreased gradually.[6] It seems that muscle damage-induced HSP70 response can reduce the CK level by exerting its cellular protective role. However, there are very few data regarding this relationship between muscle damage and the HSP response and further study is required.

One of the important mechanisms of muscle adaptations to exercise and training is muscle fiber type transition.[78,79] Different muscle fiber types differ in functional and biochemical properties; type I fibers, for example, are rich in mitochondria. HSP60 is thought to be localized in mitochondria and, therefore, the transition from fiber type II to type I can be accompanied by HSP60 increase[14] as well as increases in other mitochondrial HSPs such as GRP75.[14] It has been reported that αB-crystallin, one of the small HSPs, is expressed in tissue with high oxidative capacity[16] and it responds to muscle fiber type transition from II to I.[31] There is also evidence that HSP72 expression[4,15] is related to type I muscle fibers. In contrast, Sakuma et al.[80] reported that in a model of muscular dystrophy in the dy mouse, αB-crystallin and HSP27 were induced in the fast-twitch muscle fibers and diminished in the slow-twitch fibers. This result may be attributed to the particular disease pattern.

Because of the well-known chaperone role of HSP in facilitating protein synthesis, HSP can play an important role in the process of muscle fiber type transition and it may be possible that the above-mentioned studies describe only the protein synthesis-dependent response in particular fibers, but not a fiber-specific response. Furthermore, skeletal muscle fiber type transition is a complex process that involves intra-fiber protein degradation and synthesis, and controlled apoptic fiber decay.[78] HSPs are involved in the control of apoptosis[81–84] and cells expressing high levels of HSP70 are less sensitive to inducers of apoptosis.[82,85] However, although HSPs may be related to and play a role in muscle fiber transition, there are no data suggesting that muscle fiber transition can be indicated by the HSP response.

B.   Heat shock proteins as cytokines/messenger signals

Although HSPs may play an important role in the regulation of hormone receptor activity,[82,86] whether hormones affect the regulation of HSP expression during exercise and training is not clear. Genetic studies have demonstrated that binding of steroid receptors and some protein kinases to HSP90 is critical for their signal transducing function *in vivo*. These heterocomplexes are formed by a multiprotein chaperone machinery consisting of at least four ubiquitous proteins—HSP90, HSP70, p60, and p23.[87] Furthermore, HSP90 has been linked to activation of endothelial nitric oxide.[43] At present, a significant hormonal effect of exercise on HSP induction has only been described for estrogen. Estrogen treatment resulted in a two- to fourfold reduction in post-exercise HSP72 content in the heart, liver, lung, and red and white vastus muscles of estradiol-treated males compared with their vehicle-injected counterparts. Compared to the males, the females had significantly lower post-exercise HSP72 levels, which were not affected by estradiol supplementation.[38] We have recently observed the changes of hormones in blood plasma and the HSP70 expression in human skeletal muscle undergoing prolonged training, and no significant relation was found between HSP70 and the levels of stress hormones such as cortisol, leptin, insulin, and thyroid (unpublished data). These results suggest that the regulation of HSP70 expression in skeletal muscle is not closely linked to exercise-induced changes in plasma hormone levels.

Cytokines play an important role in the exercise-induced immune reaction and exercise-related metabolic and cellular signal transduction, and are capable of increasing HSP snythesis.[46] As mentioned, during intense exercise, levels of inflammatory cytokines (e.g., TNF-α, IL-1, and IL-8) parallel the increases in circulating leukocytes that express HSP70, ubiquitin, and other small HSPs on their surface.[8,45] With training adaptation, circulating cytokines and numbers of HSP-expressing leucocytes decrease[45] and there is an increase in muscule HSPs.[6,18] However, no clear cause-effect relationships can be drawn from these experiments.

The possibility that HSP may act itself as a cytokine in the exercise response is intriguing, with the provocative finding that exogenous HSP70 stimulates TNF-α, IL-1β, and IL-8 in human monocytes and that both CD14-dependent and calcium-dependent pathways can be activated by HSP70.[45] Furthermore, HSPs may control cytokine expression, as after heat conditioning and concomittant HSP accumulation in the liver, TNF-α-levels decrease.[88] The induction of HSP70 in tumor cells induces cytokine expression (interferon-gamma, TNF-α, IL-12) and HSP70 expression targets immature antigen-presenting cells, thereby enabling them to take up antigens.[86,89] A similar mechanism is proposed for viral antigens, whereby HSPs may act as "danger signals" in the immune conversation[86,90] such that the increased level of HSPs in skeletal muscle and other organs after training may be related to the immune response to training. Furthermore, the presence of soluble HSP70 and of circulating anti-HSP70 antibodies in the peripheral circulation of normal individuals, as well as a decrease in HSP60 and HSP70 with increasing age has been demonstrated.[91,92] The same group was also able to demonstrate increased levels of circulating HSP70 in vascular disease,[93] which would support the proposition that soluble heat shock proteins may play a regulatory role in the pathophysiological processes involving inadvertent immunorecognition.

Moseley[86] and Todryk et al.[90] recently reviewed the immune functions of HSPs and concluded that HSPs are involved in antigen presentation by antigen-presenting cells (APCs), in the targeting of natural killer cells, and modulation of the immune response. Exercise and training impose on the organism a high load of degraded proteins and cell fragments. Their subsequent activation of HSPs may then induce some unwanted autoimmune responses;[86,90] however, we must await data regarding these potential relationships between immunological responses and related HSP functions in exercise and training.

V. Perspectives

The physiological meaning of the stress protein response for physical exercise is at this time far from understood. There is a clear indication that stress proteins can be used as molecular markers of cellular stress, have important functions in the muscular repair process, and may act as cytokines and messenger signals. Further information and studies about the role of HSPs for (1) metabolism and muscle contraction; (2) fiber transformation, protein degradation, and protein synthesis; (3) hormone and cytokine receptor regulation and receptor-dependent signaling pathways; (4) the immunoresponse to exercise and training, the effects on pathophysiology of diseases like malignant tumors and the possible role in exercise-related disorders and autoimmune responses; and (5) the possible role of HSPs as cytokines are still required.

References

1. Locke, M. and Noble, E. G., Stress proteins: the exercise response, *Can. J. Appl. Physiol.*, 20, 155, 1995.
2. Hammond, G. I., Lai, K. L., and Markert, G. I., Diverse forms of stress lead to new proteins of gene expression through a common and essential pathway, *Proc. Natl. Acad. Sci., U.S.A.*, 79, 3485, 1982.
3. Locke, M., Noble, E. G., and Atkinson, B. G., Exercising mammals synthesize stress proteins, *Am. J. Physiol.*, 358, C723, 1990.
4. Locke, M., Noble, E. G., and Atkinson, B. G., Inducible isoform of HSP70 is constitutively expressed in a muscle fiber type specific pattern, *Am. J. Physiol.*, 61, C774, 1991.
5. Puntschart, A. et al., HSP70 expression in human skeletal muscle after exercise, *Acta Physiol. Scand.*, 157, 411, 1996.
6. Liu, Y. et al., Human skeletal muscle HSP70 response to training in highly trained rowers, *J. Appl. Physiol.*, 86, 101, 1999.

7. Ryan, A. J., Gisolfi, C. V., and Moseley, P. L., Synthesis of 70K stress protein by human leukocytes: effect of exercise in the heat, *J. Appl. Physiol.*, 70, 466, 1991.

8. Niess, A. M. et al., Free radicals and oxidative stress in exercise—immunological aspects, *Exerc. Immunol. Rev.*, 5:22–56, 22, 1999.

9. Fehrenbach, E. et al., Changes of HSP72-expression in leukocytes are associated with adaptation to exercise under conditions of high environmental temperature, *J. Leukoc. Biol.*, 69, 747, 2001.

10. Salo, D. C., Donovan, C. M., and Davies, K. J., HSP70 and other possible heat shock or oxidative stress proteins are induced in skeletal muscle, heart, and liver during exercise, *Free Radic. Biol. Med.*, 11, 239, 1991.

11. Thompson, H. S. et al., A single bout of eccentric exercise increases HSP27 and HSC/HSP70 in human skeletal muscle, *Acta Physiol. Scand.*, 171, 187, 2001.

12. Locke, M. et al., Shifts in type I fiber proportion in rat hindlimb muscle are accompanied by changes in HSP72 content, *Am. J. Physiol.*, 266, C1240, 1994.

13. Liu, Y. et al., Human skeletal muscle HSP70 response to physical training depends on exercise intensity, *Int. J. Sports Med.*, 21, 351, 2000.

14. Ornatsky, O. I., Connor, M. K., and Hood, D. A., Expression of stress proteins and mitochondrial chaperonins in chronically stimulated skeletal muscle, *Biochem. J.*, 311, 119, 1995.

15. Neufer, P. D. et al., Continuous contractile activity induces fiber type specific expression of HSP70 in skeletal muscle, *Am. J. Physiol.*, 271, C1828, 1996.

16. Neufer, P. D. and Benjamin, I. J., Differential expression of αB-crystallin and HSP27 in skeletal muscle during continuous contractile activity, *J. Biol. Chem.*, 271, 24089, 1996.

17. Locke, M. et al., Activation of heat-shock transcription factor in rat heart after heat shock and exercise, *Am. J. Physiol.*, 268, C1387, 1995.

18. Liu, Y. and Steinacker, J. M., Changes in skeletal muscle heat shock proteins: pathological significance, *Front Biosci.*, 6, D12, 2001.

19. Morimoto, R. I., Cells in stress: transcriptional activation of heat shock genes, *Science*, 259, 1409, 1993.

20. Li, G. C., Mivechi, N. F., and Weitzel, G., Heat shock proteins, thermotolerance, and their relevance to clinical hyperthermia, *Int. J. Hyperthermia*, 11, 459, 1995.

21. Moseley, P. L., Heat shock proteins and heat adaptation of the whole organism, *J. Appl. Physiol.*, 83, 1413, 1997.

22. Kregel, K. C. and Moseley, P. L., Differential effects of exercise and heat stress on liver HSP70 accumulation with aging, *J. Appl. Physiol.*, 80, 547, 1996.

23. Hall, D. M. et al., Aging reduces adaptive capacity and stress protein expression in the liver after heat stress, *J. Appl. Physiol.*, 89, 749, 2000.

24. Lyashko, V. et al., Comparison of the heat shock response in ethnically and ecologically different human populations, *Proc. Natl. Acad. Sci. U.S.A.*, 91, 12492, 1994.

25. Locke, M., The cellular stress response to exercise: role of stress proteins, *Exerc. Sport Sci. Rev.*, 25, 105, 1997.

26. Ding, X. Z. et al., Increases in HSF1 translocation and synthesis in human epidermoid A-431 cells: role of protein kinase C and [Ca^{2+}], *J. Investig. Med.*, 44, 144, 1996.

27. Locke, M. and Tanguay, R. M., Increased HSF activation in muscles with a high constitutive HSP70 expression, *Cell Stress Chaperones*, 1, 189, 1996.

28. Lavoie, J. G. et al., Induction of Chinese hamster HSP27 gene expression in mouse cells confers tolerance to heat shock HSP27 stabilization of the microfilament organization, *J. Biol. Chem.*, 268, 3420, 1993.

29. Quinlan, R. and van den, Ijssel, P., Fatal attraction: when chaperone turns harlot, *Nature Med.*, 5, 25, 1999.

30. Martin, J. F. et al., Small heat shock proteins and protection against ischemic injury in cardiac myocytes, *Circulation*, 96, 4343, 1997.

31. Neufer, P. D., Ordway, G. A., and Williams, R. S., Transient regulation of c-fos, alpha B-crystallin, and HSP70 in muscle during recovery from contractile activity, *Am. J. Physiol.*, 274, C341, 1998.

32. Benjamin, I. J. et al., Induction of stress proteins in cultured myogenic cells. Molecular signals for the activation of heat shock transcription factor during ischemia, *J. Clin. Invest.*, 89, 1685, 1992.

33. Sarge, K. D., Murphy, S. P., and Morimoto, R. I., Activation of heat shock gene transcription by heat shock factor 1 involves oligomerization, acquisition of DNA-binding activity, and nuclear localization and can occur in the absence of stress [published errata appear in *Mol. Cell. Biol.*, 13, 3122, 1993; *Mol. Cell. Biol.*, 13, 3838, 1993], *Mol. Cell. Biol.*, 13, 1392, 1993.

34. Fehrenbach, E. and Niess, A. M., Role of heat shock proteins in the exercise response, *Exerc. Immunol. Rev.*, 5, 57, 1999.

35. Khassaf, M. et al., Time course of responses of human skeletal muscle to oxidative stress induced by nondamaging exercise, *J. Appl. Physiol.*, 90, 1031, 2001.

36. Niess, A. M. et al., Physical exercise-induced expression of inducible nitric oxide synthase and heme oxygenase-1 in human leukocytes: effects of RRR-alpha-tocopherol supplementation, *Antioxid. Redox Signal*, 2, 113, 2000.

37. Kelly, D. A. et al., Effect of vitamin E deprivation and exercise training on induction of HSP70, *J. Appl. Physiol.*, 81, 2379, 1996.

38. Paroo, Z., Tiidus, P. M., and Noble, E. G., Estrogen attenuates HSP 72 expression in acutely exercised male rodents, *Eur. J. Appl. Physiol.*, 80, 180, 1999.

39. Gonzalez, B., Hernando, R., and Manso, R., Anabolic steroid and gender-dependent modulation of cytosolic HSP70s in fast- and slow-twitch skeletal muscle, *J. Steroid Biochem. Mol. Biol.*, 74, 63, 2000.

40. Knoblauch, R. and Garabedian, M. J., Role for HSP90-associated cochaperone p23 in estrogen receptor signal transduction, *Mol. Cell. Biol.*, 19, 3748, 1999.

41. Caruso, J. A., Laird, D. W., and Batist, G., Role of HSP90 in mediating cross-talk between the estrogen receptor and the Ah receptor signal transduction pathways, *Biochem. Pharmacol.*, 58, 1395, 1999.

42. Grammatikakis, N. et al., p50(cdc37) acting in concert with HSP90 is required for Raf-1 function, *Mol. Cell. Biol.*, 19, 1661, 1999.

43. Garcia-Cardena, G. et al., Dynamic activation of endothelial nitric oxide synthase by HSP90, *Nature*, 392, 821, 1998.

44. Arrigo, A.-P., Tumor necrosis factor induces the rapid phosphorylation of the mammalian heat shock protein HSP28, *Mol. Cell. Biol.*, 10, 12176, 2001.

45. Fehrenbach, E. et al., Transcriptional and translational regulation of heat shock proteins in leukocytes of endurance runners, *J. Appl. Physiol.*, 89, 704, 2000.

46. Asea, A. et al., HSP70 stimulates cytokine production through a CD14-dependant pathway, demonstrating its dual role as a chaperone and cytokine, *Nat. Med.*, 6, 435, 2000.

47. Morimoto, R. I., Heat shock: the role of transient inducible responses in cell damage, transformation, and differentiation, *Cancer Cells*, 3, 295, 1991.

48. Skidmore, R. et al., HSP70 induction during exercise and heat stress in rats: role of internal temperature, *Am. J. Physiol.*, 268, R92, 1995.

49. Febbraio, M. A. and Koukoulas, I., HSP72 gene expression progressively increases in human skeletal muscle during prolonged, exhaustive exercise, *J. Appl. Physiol.*, 89, 1055, 2000.

50. Flanagan, S. W. et al., Tissue-specific HSP70 response in animals undergoing heat stress, *Am. J. Physiol.*, 268, R28, 1995.

51. Comini, L. et al., Right heart failure chronically stimulated heat shock protein 72 in heart and liver but not in other tissues, *Cardiovasc. Res.*, 31, 882, 1996.

52. Manzerra, M., Rush, S. J., and Brown, I. R., Tissue-specific differences in heat shock protein HSC70 and HSP70 in the control and hyperthermic rabbit, *J. Cell. Physiol.*, 170, 130, 1997.

53. Samelman, T. R., Heat shock protein expression is increased in cardiac and skeletal muscles of Fischer 344 rats after endurance training, *Exp. Physiol.*, 85, 92, 2000.

54. Powers, S. K. et al., Exercise training improves myocardial tolerance to *in vivo* ischemia in the rat, *Am. J. Physiol.*, 275, R1468, 1998.

55. Bonventre, J. V., Mediators of ischemic renal injury, *Ann. Rev. Med.*, 39, 531, 1988.

56. Linquist, S. and Craig, E. A., The heat-shock proteins, *Annu. Rev. Genet.*, 22, 631, 1988.

57. Knowlton, A. A., Brecher, P., and Apstein, C. S., Rapid expression of heat shock protein in the rabbit after brief cardiac ischemia, *J. Clin. Invest.*, 87, 139, 1991.

58. Dillmann, W. H. and Mestril, R., Heat shock proteins in myocardial stress, *Z. Kardiol.*, 84 (Suppl. 4), 87, 1995.

59. Abe, K. et al., Preferential expression of HSC70 heat shock mRNA in gerbil heart after transient brain ischemia, *J. Mol. Cell. Cardiol.*, 25, 1131, 1993.

60. Gingalewski, C. et al., Distinct expression of heat shock and acute phase genes during hepatic ischemia-reperfusion, *Am. J. Physiol.*, 271, R634, 1996.

61. Aufricht, C. et al., ATP releases HSP-72 from protein aggregates after renal ischemia, *Am. J. Physiol.*, 274, F268, 1998.

62. Smoyer, W. E. et al., Ischemic acute renal failure induces different expression of small heat shock proteins, *J. Am. Soc. Nephrol.*, 11, 211, 2000.

63. Liauw, S. K. et al., Sequential ischemia/reperfusion results in contralateral skeletal muscle salvage, *Am. J. Physiol.*, 270, H1407, 1996.

64. Lepore, D. A. et al., Prior heat stress improves survival of ischemic-reperfused skeletal muscle *in vivo*, *Muscle Nerve*, 23, 1847, 2000.

65. Ecochard, L. et al., Skeletal muscle HSP72 level during endurance training: influence of peripheral arterial insufficiency, *Pflügers Arch.*, 440, 918, 2000.

66. Ku, Z. et al., Decreased polysomal HSP-70 may slow polypeptide elongation during skeletal muscle atrophy, *Am. J. Physiol.*, 268, C1369, 1995.

67. Naito, H. et al., Heat stress attenuates skeletal muscle atrophy in hindlimb-unweighted rats, *J. Appl. Physiol.*, 88, 359, 2000.

68. Yamakuchi, M. et al., Type I muscle atrophy caused by microgravity-induced decrease of myocyte enhancer factor 2C (MEF2C) protein expression, *FEBS Lett.*, 477, 135, 2000.

69. Maiello, M. et al., Basal synthesis of heat shock protein 70 increases with age in rat kidneys, *Gerontology*, 44, 15, 2001.

70. Kregel, K. C. et al., HSP70 accumulation in tissues of heat-stressed rats is blunted with advancing age, *J. Appl. Physiol.*, 79, 1673, 1995.

71. Locke, M. and Tanguay, R. M., Diminished heat shock response in the aged myocardium, *Cell Stress Chaperones,* 1, 251, 1996.

72. Locke, M., Heat shock transcription factor activation and HSP72 accumulation in aged skeletal muscle, *Cell Stress Chaperones,* 5, 45, 2000.

73. Kilgore, J. L. et al., Stress protein induction in skeletal muscle: comparison of laboratory models to naturally occurring hypertrophy, *J. Appl. Physiol.,* 76, 598, 1994.

74. Izumo, S., Nadal-Ginard, B., and Mahdavi, V., Protooncogene induction and reprogramming of cardiac gene expression produced by pressure overload, *Proc. Natl. Acad. Sci. U.S.A.,* 85, 339, 1988.

75. Lehmann, M. et al., Selected parameters and mechanisms of peripheral and central fatigue and regeneration in overtrained athletes, in Lehmann, M. et al., Eds., *Overload, Performance Incompetence and Regeneration in Sport,* Kluwer Academic/Plenum Pubishers, New York, 1999, 7–26.

76. Lehmann, M. J. et al., Training and overtraining: an overview and experimental results in endurance sports, *J. Sports Med. Phys. Fitness,* 37, 7, 1997.

77. Nosek, T. M. et al., Functional properties of skeletal muscle from transgenic animals with upregulated heat shock protein 70, *Physiol. Genomics,* 4, 25, 2000.

78. Pette, D. and Staron, R. S., Mammalian skeletal muscle fiber type transitions, *Int. Rev. Cytol.,* 170, 143, 1997.

79. Schiffiano, S. and Reggiani, C., Molecular diversity of myofibrillar proteins: gene regulation and functional significance, *Physiol. Rev.,* 76, 371, 1996.

80. Sakuma, K. K. et al., Pathological changes in levels of three small stress proteins, αB-crystallin, HSP27 and P20, in the hindlimb of dy mouse, *Biochem. Biophys. Acta,* 1406, 162, 1998.

81. Beere, H. M. and Green, D. R., Stress management—heat shock protein-70 and the regulation of apoptosis, *Trends Cell. Biol.,* 11, 6, 2001.

82. Brar, B. K. et al., Heat shock proteins delivered with a virus vector can protect cardiac cells against apoptosis as well as against thermal or hypoxic stress, *J. Mol. Cell. Cardiol.,* 31, 135, 1999.

83. Moseley, P., Stress proteins and the immune response, *Immunopharmacology,* 48, 299, 2000.

84. Mosser, D. D. et al., Role of the human heat shock protein HSP70 in protection against stress-induced apoptosis, *Mol. Cell. Biol.,* 17, 5317, 1997.

85. Mosser, D. D. et al., The chaperone function of HSP70 is required for protection against stress-induced apoptosis, *Mol. Cell. Biol,* 20, 7146, 2000.

86. Moseley, P. L., Exercise, stress, and the immune conversation, *Exerc. Sport Sci. Rev.,* 28, 128, 2000.

87. Pratt, W. B., The HSP90-based chaperone system: involvement in signal transduction from a variety of hormone and growth factor receptors, *Proc. Soc. Exp. Biol. Med.,* 217, 420, 1998.

88. Kluger, M. J. et al., Effect of heat stress on LPS-induced fever and tumor necrosis factor, *Am. J. Physiol.,* 273, R858, 1997.

89. Todryk, S. M. et al., Heat shock protein 70 induced during tumor cell killing induces Th1 cytokines and targets immature dendritic cell precursors to enhance antigen uptake, *J. Immunol.,* 163, 1398, 1999.

90. Todryk, S. M. et al., Heat shock proteins refine the danger theory, *Immunology,* 99, 334, 2000.

91. Pockley, A. G., Shepherd, J., and Corton, J. M., Detection of heat shock protein 70 (HSP70) and anti-HSP70 antibodies in the serum of normal individuals, *Immunol. Invest.*, 27, 367, 1998.
92. Rea, I. M., McNerlan, S., and Pockley, A. G., Serum heat shock protein and anti-heat shock protein antibody levels in aging, *Exp. Gerontol.*, 36, 341, 2001.
93. Wright, B. H. et al., Elevated levels of circulating heat shock protein 70 (HSP70) in peripheral and renal vascular disease, *Heart Vessels*, 15, 18, 2000.

Index

A

Accessory proteins, 51
N-Acetylcysteine, 129
Acidosis, 199–200
Actin, 50
Actinomycin D, 28
Activator protein-1, 127
Acute exercise, *See* Exercise
Adenovirus, 189, 190
Aging, 146
    circulating HSP levels and, 211
    HSP accumulation and, 208
    muscle deterioration, 141, 145–146
Alarm reaction stage, 2
Alkyl-lysophospholipids, 184
Allopurinol, 129
Alpha-beta-crystallin, *See* αβ-Crystallin
3-Aminotriazole, 105
Anabolic steroids, 59–60
Anoxia, 143
Antigen-presenting cells, 185, 188,
    211
Antioxidant enzymes, *See specific*
    enzymes
Antisense oligonucleotides, 170
Apoptosis, 48, 49, 59, 187, 188, 209
    transfection model, 104
Atherosclerotic lesion, 47
ATPase, 51, 80–81
ATP depletion, 199–200
ATP intracellular concentrations, 51
Auto-immune response, 186, *See*
    Immune response
Avian adenovirus, 189

B

Bag-1, 51
Beet yellow virus, 190
Beta-carotene, 142
Biliverdin, 47
Body temperature, exercise-induced
    HSP response, 168, 198–199
Brownian ratchet model, 156
Butylated hydroxytoluene, 129

C

Cadmium sulfate, 30
Caffeine sensitivity, 209
Calcineurin, 59, 83
Calcium, 59, 107, 108, 199
Calcium channels, 116
Calmodulin, 59
Canavanine, 28, 30
Carbon monoxide, 47
Cardiac muscle
    cardioprotective effects of HSPs, *See*
        Cardioprotection
    HSP expression, 56–57, 86
    mitochondrial biogenesis, 151
    reactive oxygen species, 125
Cardiolipin, 157
Cardioprotection, 58, 97–117, 143, *See*
    also Ischemia-reperfusion
    coronary vasculature dysfunction,
        114–116
    exercise and, 106–114
        acute exercise studies, 58, 111–114
        antioxidant enzymes, 109–114